BIGFOOT

Hidden in Not So Plain Sight

By Daniel Reidmiller

First Edition: January 27, 2025

Printed in the United States of America

ISBN: 979-8-3485-5368-5

Foreword

As of January 5, 2025, I have walked this planet for 69 years, and for 56 of those years, I've been captivated by the thrilling mysteries of cryptozoology. It has been a journey filled with discovery, fueled by an insatiable curiosity and a belief that the world still holds secrets waiting to be uncovered.

Bigfoot/Sasquatch has always fascinated me, sitting just a notch above Nessie and other cryptids. This elusive and enigmatic creature represents the ultimate frontier in our understanding of the natural world.

Over the years, I've immersed myself in countless books about cryptids, absorbed the compelling narratives of television shows like the 1970s classic *In Search Of* series, more recently, *Expedition Bigfoot,* and listening to *Cliff Barackman and James "Bobo" Fay* with their *Bigfoot and Beyond* podcast. These programs have not only increased my knowledge but also tossed logs onto my fire for further examination into all things cryptid, particularly Bigfoot.

I owe much of my inspiration to the giants in this field, those who dared to seek answers where others scoffed or hesitated. Grover Krantz, with his scientific rigor; Ivan T. Sanderson, the consummate storyteller, and explorer; Loren Coleman, whose works guided my early forays into cryptozoology; and modern icons like Cliff Barackman, James "Bobo" Fay, and Dr. Jeff Meldrum, who've brought the search for Bigfoot to new audiences with dedication and humor.

Special acknowledgment must go to two remarkable women: Dr. Jane Goodall, whose groundbreaking work in primatology has inspired cryptozoologists to approach Bigfoot as a biological possibility rather than mere folklore, and Dr. Mireya Mayor, whose scientific expertise and fieldwork reminds us that there is still so much to discover in this world.

These individuals and countless others have shaped my understanding and appreciation for the vastness of what we know—and, more importantly, what we *don't* know. Their work has taught me that pursuing the unknown is as much about the journey as the destination.

So this book is a culmination of the things I have learned over my life, put down in a reasonably simple way, to perhaps nudge those aged 13 and older to engage in a subject that has fed my curiosity and imagination for over half a century. It sees Bigfoot as a creature and symbol of mystery, resilience, and the untamed wild.

Whether you are a steadfast believer, an open-minded skeptic, or simply curious, I hope this book inspires in you the same sense of wonder that has driven my journey.

To the explorers, the dreamers, and the seekers: may the unknown always call to you.

Please, enjoy.

Chapter 1

The Bigfoot legend has captivated the imaginations of countless individuals around the world. This enigmatic creature, often described as a large, bipedal, ape-like figure, has been the subject of folklore, eyewitness accounts, and extensive speculation for centuries.

Known by various names—Sasquatch in North America, Yeti in the Himalayas, and Yowie in Australia—the creature is known by cultures worldwide.

In North America, stories of Sasquatch predate European colonization. Indigenous peoples told tales of a giant, hairy "wild man" who lived in the forests, a being both feared and revered. These oral traditions laid the foundation for the modern-day fascination with Bigfoot, merging myth and mystery in a way that continues to capture public interest.

Bigfoot first entered mainstream consciousness in the late 1950s when large footprints were discovered near Bluff Creek, California. Measuring over 16 inches in length, the tracks were unlike anything previously seen and ignited a media frenzy. Newspapers dubbed the creature "Bigfoot," a term that quickly caught on. The discovery of the footprints coincided with a surge in reported sightings, cementing Bigfoot as a cultural phenomenon.

The Patterson-Gimlin film, recorded in 1967, further fueled interest in Bigfoot. Shot in Northern California, the footage shows a large, hairy figure walking upright through the forest. To this day, the film remains one of the most hotly debated pieces of evidence, with supporters hailing it as proof of Bigfoot's existence and skeptics dismissing it as an elaborate hoax.

Despite the lack of definitive proof, Bigfoot enthusiasts continue their quest for answers. Researchers and amateur investigators scour forests, set up trail cameras, and analyze audio recordings, hoping to uncover evidence of the elusive creature. Some focus on physical evidence, such as footprints, hair, or scat, while others investigate alleged vocalizations or strange, unexplained structures in the woods. Each piece of evidence, no matter how small, contributes to the enduring allure of Bigfoot.

But Bigfoot’s appeal isn’t limited to the realm of research. The creature has become a cultural icon in movies, television shows, and advertising campaigns. From the terrifying figure depicted in "The Legend of Boggy Creek" to the lovable character of "Harry and the

Hendersons," Bigfoot has cemented its place in the popular imagination.

The ongoing search for Bigfoot speaks to something deeper within the human psyche—a longing for mystery and the unknown. In a world where so much has been mapped, studied, and explained, Bigfoot represents the possibility that there are still wild places and hidden creatures waiting to be discovered. Whether a living creature, a cultural symbol, or a product of folklore, Bigfoot endures as one of the most intriguing mysteries of our time.

Chapter 2

Bigfoot sightings have been reported for centuries, yet the mystery surrounding the creature remains as strong as ever. Accounts range from fleeting glimpses in the forest to detailed encounters, with witnesses describing a massive, bipedal figure covered in hair, standing anywhere from seven to nine feet tall. These stories have captivated people from all walks of life—hunters, hikers, and ordinary citizens alike. But what drives these sightings? Are they genuine encounters with an unknown species, or do they stem from something else entirely?

One of the earliest recorded accounts of Bigfoot-like creatures comes from indigenous North American folklore. Many tribes spoke of large, hairy beings that lived in the forests and mountains. The Coast Salish referred to them as "Sasquatch," while other groups had their own names and interpretations of these creatures. Some saw them as protectors of the wilderness, while others described them as malevolent spirits to be avoided. These tales laid the foundation for the modern Bigfoot legend, blending myth and reality in a way that continues to resonate today.

In the 19th and early 20th centuries, stories of wild men and mysterious creatures began appearing in newspapers and journals. Settlers and explorers reported encounters with large, hairy beings, often describing them as "ape-men" or "giant wild men of the woods." While many dismissed these stories as tall tales or misidentifications of known animals, others saw them as evidence of something truly extraordinary.

The 1958 discovery of massive footprints near Bluff Creek, California, brought Bigfoot into the national spotlight. Workers at a logging site found tracks measuring 16 inches long and eight inches wide, with a stride length that suggested a creature of immense size and strength. The story made headlines, and the term "Bigfoot" was coined. While these tracks were later revealed to be part of a hoax orchestrated by Ray Wallace, they sparked a wave of interest in Bigfoot that has never entirely subsided.

Perhaps the most famous piece of evidence in the Bigfoot saga is the 1967 Patterson-Gimlin film. Shot in Bluff Creek, the grainy footage shows a large, upright figure walking through a clearing before disappearing into the forest. The figure's movements, proportions, and muscle tone have been analyzed extensively, with some experts claiming it couldn't have been a man in a suit. Skeptics, however, argue that the film lacks corroborating evidence and is likely a hoax. Despite decades of debate, the film remains one of the most iconic images associated with Bigfoot.

Beyond these high-profile cases, thousands of Bigfoot sightings have been reported across the United States, particularly in the Pacific Northwest, where dense forests provide the perfect cover for an elusive creature. Witnesses often describe similar features: a tall, muscular build, long arms, and a face that combines ape-like and human characteristics. Many also report a pungent smell, often described as a mix of wet dogs and rotting vegetation.

Audio evidence has also played a role in the Bigfoot mystery. Recordings of strange howls, screams, and wood-knocking sounds have been attributed to the creature, though skeptics argue that these noises could be the calls of known animals or natural phenomena. One of the most famous recordings, the "Sierra Sounds," captured in the 1970s by Ron Morehead and Al Berry, features vocalizations that some believe exhibit language-like patterns. Linguists and audio experts remain divided on the authenticity of these recordings, but they continue to fascinate Bigfoot enthusiasts.

Despite the lack of definitive proof, the sheer volume of eyewitness accounts and reported evidence keeps the Bigfoot legend alive. Skeptics point to misidentifications, hoaxes, and psychological factors as explanations for sightings, but believers argue that the consistency across reports suggests a real, undiscovered species. Whether viewed as a cryptid, a myth, or a cultural phenomenon, Bigfoot continues to ignite debate and curiosity around the world.

Chapter 3

The search for Bigfoot isn't just about proving the creature's existence; it's also a journey into the unknown. For many, the mystery of Bigfoot symbolizes humanity's deep-rooted curiosity about the natural world and the possibility that secrets are still waiting to be discovered. This chapter explores the methods and challenges of Bigfoot research and the passion and determination of those who dedicate their lives to pursuing this elusive legend.

One of the most popular tools used in Bigfoot research is the trail camera. These motion-activated devices, commonly employed by hunters and wildlife enthusiasts, have become a staple for cryptid investigators. Placed in areas with reported sightings, trail cameras capture images or video of anything that moves within their range. While these cameras have successfully documented bears, deer, and other wildlife, they have yet to capture indisputable evidence of Bigfoot. Some researchers speculate that Bigfoot may be intelligent enough to avoid cameras, detecting the faint hum or infrared light they emit. Others suggest that the creature's rarity and stealth explain the lack of conclusive footage.

Audio recordings also play a significant role in the search for Bigfoot. Researchers often deploy high-quality audio recorders in remote areas, hoping to capture vocalizations attributed to the creature. Strange howls, screams, and wood-knocking sounds have been documented, with some recordings analyzed for their frequency, tone, and linguistic patterns. One notable example is the "Sierra Sounds," a series of audio recordings made in the 1970s. Supporters believe these sounds indicate an unknown species communicating in a complex, language-like manner. Still, skeptics argue that they could be the calls of known animals or even elaborate hoaxes.

Footprint analysis is another cornerstone of Bigfoot research. Casts of alleged Bigfoot tracks are carefully examined for details such as dermal ridges, toe impressions, and stride length. Some of these casts show features that researchers claim would be difficult to fake, such as intricate patterns resembling human fingerprints. However, many footprints are dismissed as hoaxes, misidentifications, or the result of natural processes like erosion. The challenge lies in distinguishing genuine evidence from fabricated or ambiguous findings.

In recent years, advances in technology have opened new doors for Bigfoot research. Drones equipped with high-definition cameras and thermal imaging allow researchers to survey large wilderness areas from above, covering terrain that would be difficult or dangerous to navigate on foot. Thermal cameras, in particular, can detect heat signatures, potentially revealing the presence of a large, warm-blooded creature hiding in the dense forest. While these tools have not produced definitive evidence, they represent a significant leap forward in the search for Bigfoot.

One of the most promising developments in wildlife research is environmental DNA (eDNA) sampling. This technique involves collecting DNA from the environment, such as soil, water, or air, rather than from the creature itself. Animals leave traces of DNA through skin cells, hair, saliva, or waste, and scientists can analyze these samples to identify the species present in a given area. eDNA has already been used to detect rare or elusive species, and some Bigfoot researchers believe it could provide the breakthrough they've been seeking. However, no confirmed Bigfoot DNA has been identified to date.

Field expeditions are another critical component of Bigfoot research. Researchers and enthusiasts often venture into areas with a history of sightings equipped with cameras, audio recorders, and other tools.

These expeditions require careful planning and a willingness to endure harsh conditions, as many reported sightings occur in remote, densely forested regions. While these trips often yield fascinating anecdotes and intriguing clues, they rarely produce concrete evidence.

One of the biggest challenges facing Bigfoot researchers is the prevalence of hoaxes. Over the years, numerous individuals have fabricated evidence, from fake footprints to staged sightings, in an attempt to gain attention or profit from the phenomenon. These hoaxes undermine the credibility of genuine research and make it harder for the public and scientific community to take the subject seriously. Researchers must remain vigilant, carefully scrutinizing all evidence to separate the real from the fake.

Despite the obstacles, the search for Bigfoot continues to captivate the imagination of those involved. For many researchers, the quest is about more than just finding a cryptid; it's a way to connect with the natural world, challenge the boundaries of knowledge, and explore the mysteries of existence.

Whether or not Bigfoot is ever found, the passion and curiosity that drive the search are a testament to humanity's unyielding desire to understand the unknown.

Chapter 4

Bigfoot has been a subject of fascination for centuries, with stories of large, hairy, human-like creatures appearing in the folklore of cultures around the world. In North America, the legend of Bigfoot is deeply intertwined with Indigenous traditions, where these creatures are often seen as guardians of the forest, spiritual beings, or even ancestors. This chapter delves into Bigfoot lore's historical and cultural roots, exploring how these stories have shaped modern perceptions of the creature.

Indigenous tribes across North America have long spoken of creatures resembling what we now call Bigfoot. In the Pacific Northwest, the Coast Salish people refer to Sasquatch as a protector of the wilderness, a being that should be respected and left undisturbed. Similarly, the Lakota Sioux tell stories of the Chiye-Tanka, a "big elder brother" who watches over the land. These tales often portray Bigfoot as a powerful and mysterious figure, embodying the untamed forces of nature.

One of the most intriguing aspects of Indigenous Bigfoot stories is their consistency across diverse cultures. Tribes separated by vast distances and with little interaction often describe creatures with similar characteristics: large, bipedal, covered in hair, and elusive. This raises fascinating questions about the origins of these legends. Were they inspired by actual encounters with unknown creatures, or do they reflect a shared human archetype, a deep-seated need to personify the wilderness?

European settlers in North America added their own interpretations to the Bigfoot legend. Early explorers and pioneers reported sightings of wild men in the forests, describing them as enormous, hairy beings with a distinctly human-like appearance. These accounts often blurred the line between myth and reality, with some settlers viewing the creatures as literal beings and others seeing them as symbols of the untamed frontier.

In the 19th and early 20th centuries, newspapers and magazines became fascinated with Bigfoot. Stories of mysterious footprints, strange howls, and fleeting glimpses of large, shadowy figures captivated readers, fueling public curiosity. Some articles treated the creature as a legitimate zoological mystery, while others leaned into sensationalism, portraying Bigfoot as a monster lurking in the wilderness. This media attention helped solidify Bigfoot's place in popular culture, laying the groundwork for the modern phenomenon.

The release of the Patterson-Gimlin film in 1967 marked a turning point in the history of Bigfoot lore. Captured in the forests of Northern California, the grainy footage depicts a large, hairy creature walking upright along a riverbank. The film quickly became one of the most famous and controversial pieces of evidence for Bigfoot's existence, sparking debates that continue to this day. For believers, the film represents undeniable proof of the creature's reality. For skeptics, it is a cleverly executed hoax or a case of misidentification.

In addition to the Patterson-Gimlin film, numerous sightings and encounters have been reported over the years, each adding to the rich tapestry of Bigfoot lore. Some witnesses describe seeing the creature crossing roads at night, while others recount finding massive footprints in the snow. Hunters, hikers, and campers have shared stories of being watched by something in the forest, an experience that leaves them with a sense of awe and unease.

While skeptics argue that these sightings are the result of hoaxes, misidentifications, or overactive imaginations, believers point to the consistency of the reports as evidence that something real is out there. The descriptions of Bigfoot vary in detail but often share common themes: a massive, upright figure with long arms, broad shoulders, and a face that is both human-like and ape-like. These similarities have led some researchers to speculate that Bigfoot could be a relic population of a prehistoric hominid, such as Gigantopithecus or Paranthropus.

The cultural significance of Bigfoot extends beyond North America. Stories of large, hairy, human-like creatures appear in the folklore of many cultures around the world. In the Himalayas, the Yeti, or "Abominable Snowman," is said to inhabit the snowy peaks and remote valleys, while in Australia, the Yowie is described as a giant, hairy being that roams the bushland. These international legends share striking similarities with the Bigfoot stories of North America, suggesting a universal fascination with the idea of wild men lurking in the wilderness.

Some anthropologists believe that these legends may have a basis in humanity's collective memory. For much of our history, Homo sapiens coexisted with other hominid species, such as Neanderthals and Denisovans. Encounters with these species could have inspired stories that were passed down through generations, eventually evolving into the myths and legends we know today. Others argue that these stories reflect a more profound psychological need to grapple with the mysteries of the natural world, personifying its untamed and unpredictable forces.

Whatever their origins, the stories of Bigfoot and its global counterparts continue to captivate the imagination. They remind us that, even in an age of satellite imagery and scientific breakthroughs, there are still mysteries that elude our understanding. The legend of Bigfoot is more than just a tale of a cryptid; it is a reflection of humanity's enduring curiosity and our desire to connect with the unknown.

Chapter 5

For many, the idea of Bigfoot is tied to the forests of the Pacific Northwest, with its towering trees and misty, remote wilderness. However, sightings and encounters with mysterious, hairy creatures have been reported across North America, from the swamps of Florida to the plains of the Midwest. This chapter explores some of the regional variations in Bigfoot sightings and examines how local environments and cultures influence the way the creature is perceived.

In the Pacific Northwest, Bigfoot is often depicted as a solitary guardian of the deep forest, a creature of immense size and strength. The dense, coniferous forests and rugged terrain of this region provide the perfect backdrop for the legend, offering ample cover for a creature that wishes to remain hidden. Many of the most famous Bigfoot sightings, including the

Patterson-Gimlin film, have occurred in this area, cementing its reputation as the epicenter of Bigfoot activity.

Moving south to California, Bigfoot lore begins to take on new dimensions. The forests of Northern California, where the Patterson-Gimlin footage was filmed, are home to numerous sightings and footprint discoveries. In the Sierra Nevada Mountains, hikers and campers often report hearing strange vocalizations, wood knocks, and other unexplained sounds. These encounters are sometimes accompanied by the discovery of large, human-like footprints, which have been cast and studied by researchers.

In the swamps and wetlands of the southeastern United States, the legend of Bigfoot takes on a different form. Here, the creature is often referred to as the Skunk Ape, named for its reported foul odor. Sightings of the

Skunk Ape are most common in Florida, particularly in the Everglades and other swampy regions. Witnesses describe a smaller, more agile creature than the Bigfoot of the Pacific Northwest, often covered in reddish-brown hair and with a hunched posture. The Skunk Ape's reputation is one of mystery and mischief, with some stories describing it as a trickster figure that steals food and evades capture with cunning.

In the Midwest, Bigfoot encounters often occur in rural areas and along riverbanks. These sightings tend to describe a creature similar to its Pacific Northwest counterpart, though often portrayed as more curious than intimidating. In Ohio, the creature is known as the Grassman, named for its alleged habit of building nests out of grass and vegetation. The Grassman has been the subject of numerous investigations, with researchers examining reported tracks, hair samples, and eyewitness accounts.

The Appalachians, with their sprawling forests and rugged mountains, are another hotspot for Bigfoot activity. In this region, the creature is sometimes called the Wood Booger, a term that reflects the playful and mischievous tone of many Appalachian stories. Sightings in the Appalachians often involve vocalizations, such as long, mournful howls or eerie screams that echo through the valleys. Locals sometimes attribute these sounds to Bigfoot, adding to the mystique of the region.

The plains and deserts of the American West might seem like unlikely places for Bigfoot sightings, but reports persist even in these open landscapes. In states like Oklahoma and Texas, the creature is sometimes referred to as the Boggy Creek Monster, named after the famous 1972 film "The Legend of Boggy Creek." Sightings in these areas often describe a more aggressive creature, one that is less shy about approaching human settlements. Stories of livestock being harassed or attacked are not uncommon, fueling the image of Bigfoot as a territorial predator.

In the northern United States and Canada, the creature is often associated with the wilderness and the First Nations' legends of the Sasquatch. In these regions, Bigfoot is sometimes seen as a spiritual being, a guardian of the forest who exists on the boundary between the natural and the supernatural. Many Indigenous communities view the creature with a mixture of respect and caution, seeing it as a reminder of humanity's connection to the natural world.

One of the most intriguing aspects of Bigfoot sightings is their consistency across regions, despite variations in local lore and terminology. Whether it's the Skunk Ape of Florida, the Grassman of Ohio, or the Sasquatch of British Columbia, the descriptions often share common elements: a tall, hairy, bipedal creature with immense strength and an uncanny ability to avoid detection. This consistency has led some researchers to propose that Bigfoot is a real, biological species that has adapted to different environments across North America.

Skeptics, however, argue that regional variations in Bigfoot lore are evidence of cultural storytelling rather than a real creature. They point out that many Bigfoot sightings can be attributed to misidentifications of known animals, such as bear or elk, or to hoaxes and overactive imaginations. The differences in descriptions, such as the size and behavior of the creature, are often seen as further evidence that Bigfoot is a product of local folklore rather than a single, consistent phenomenon.

Despite the skepticism, the regional diversity of Bigfoot lore adds a fascinating layer to the legend. Each variation reflects the unique character of the landscape and culture in which it exists, offering a glimpse into the way humans interpret the mysteries of their surroundings. Whether as a symbol of the wild, a guardian of the forest, or a playful trickster, Bigfoot continues to inspire curiosity and wonder across the continent.

Chapter 6

The search for Bigfoot is not confined to scientists or professional researchers. Over the years, thousands of hunters, hikers, campers, and everyday individuals have reported encounters with the elusive creature. These firsthand accounts form the backbone of Bigfoot lore, offering tantalizing glimpses into what might be lurking in the deep wilderness. While the stories vary in detail, they often share common themes: a massive, hairy figure, a powerful presence, and an experience that leaves witnesses questioning what they thought they knew about the world.

Some of the most compelling stories come from those who spend significant time in the wilderness. Hunters, in particular, are often familiar with the animals and sounds of the forest, making their accounts of unusual encounters stand out. One such story comes from a bowhunter in the mountains of Northern California, who had an experience he would never forget.

The hunter had been tracking deer through a dense section of forest, moving quietly so as not to alert his prey. As he reached a clearing, he noticed an eerie silence settling over the woods. The birds had stopped chirping, and the usual rustling of small animals in the underbrush was absent. It was as if the forest itself was holding its breath. He paused, listening, when suddenly he heard a series of loud, rhythmic knocks echoing through the trees. The sound was unlike anything he had ever heard before, too deep and deliberate to be caused by wind or animals. It sounded like someone—or something—was striking wood against wood.

Curious but cautious, the hunter moved closer to the source of the sound. As he approached the ridge, he caught a glimpse of movement. A large, dark figure was standing in the trees about fifty yards away, partially obscured by the foliage. It stood upright, easily seven or eight feet tall, with broad shoulders and long arms that hung almost to its knees. The creature seemed to be watching him, swaying slightly as if assessing the situation. For a moment, the two locked eyes, and the hunter felt a deep, instinctual fear welling up inside him. This was not an animal he was familiar with, and every fiber of his being told him to back away slowly.

Without breaking eye contact, the creature turned and disappeared into the forest, moving with surprising grace for something so large. The hunter stood frozen for a moment, heart pounding, before quickly retreating to his truck. He later reported the encounter to local Bigfoot researchers, who investigated the area but found no physical evidence. Still, the hunter remained convinced that what he had seen was not a bear or any other known animal. To this day, he refuses to return to that part of the woods.

Encounters like this one are not uncommon in areas with dense forest cover and limited human activity. The Pacific Northwest, the Sierra Nevada mountains, and the Appalachian region are all hotspots for Bigfoot sightings. Many witnesses report similar experiences: unusual silence in the forest, followed by strange knocks or vocalizations, and then a fleeting glimpse of a massive, bipedal figure.

One of the most chilling aspects of these encounters is the vocalizations attributed to Bigfoot. Witnesses often describe hearing howls, screams, and whoops that echo through the night, unlike the sounds of any known animal. In some cases, these vocalizations are accompanied by other strange phenomena, such as wood-knocking (the sound of large sticks or branches being struck against trees) or the feeling of being watched. For those who experience it, the combination of these sensory inputs is enough to make them question their safety in the woods.

One of the most famous recorded Bigfoot vocalizations is known as the "Sierra Sounds," captured in the early 1970s by Ron Morehead and Al Berry in the Sierra Nevada mountains of California. The two men had been camping in a remote area when they began hearing strange, loud howls and screams that seemed to come from multiple directions. Over several nights, they recorded hours of audio, capturing what many believe to be the vocalizations of one or more Bigfoot creatures.

The Sierra Sounds are unsettling, to say the least. The vocalizations are deep, guttural, and eerily human-like, but with a power and range that far exceeds what a human voice could produce. Some of the sounds resemble whoops or howls, while others sound more like chatter, almost as if the creatures are communicating with each other.

Linguists and audio experts have analyzed the recordings, with some suggesting that the sounds exhibit patterns of language, while others believe they are simply the calls of unknown or misidentified animals. Despite extensive analysis, the origin of the Sierra Sounds remains a mystery.

Another account comes from a group of campers in Washington's Mount Rainier National Park. After setting up camp near a remote lake, the group spent the evening fishing and enjoying the peaceful surroundings. As night fell, they built a fire and settled in for the evening, but soon noticed an odd smell—musky and pungent, like a mix of wet dog and rotting vegetation. The smell grew stronger as the night went on, and one of the campers remarked that it felt as though something was watching them from the tree line.

As the group sat around the fire, they suddenly heard heavy footsteps approaching from the darkness. The sound was unmistakable: large, deliberate steps, crunching leaves and snapping branches. They shined their flashlights toward the sound but saw nothing. The footsteps stopped just outside the range of their lights, and for several tense minutes, the group sat in silence, listening as whatever was out there paced back and forth in the shadows.

Just as suddenly as it had arrived, the presence seemed to retreat, leaving the group shaken but unharmed. The next morning, they found large footprints in the soft earth around their campsite, far larger than any human could make.

Though they tried to rationalize the encounter—perhaps it was a bear or another animal—the size and shape of the prints, along with the strange behavior, left them questioning what they had experienced.

These stories, while anecdotal, form an important part of the Bigfoot phenomenon. For every physical footprint or blurry photograph, there are dozens of encounters that leave witnesses with more questions than answers. What makes these experiences so compelling is the consistency across accounts. People from different walks of life, in different regions, describe seeing or hearing similar things: a large, bipedal creature, strange vocalizations, and an overwhelming sense that they are in the presence of something beyond the ordinary.

While skeptics argue that these encounters are the result of misidentifications, hoaxes, or overactive imaginations, believers see them as pieces of a larger puzzle. Together, these stories suggest that there is something out there—something big, intelligent, and elusive—that has yet to be fully understood. Whether it's a new species of primate, an undiscovered hominid, or something else entirely, the search for answers continues, driven by the accounts of those who have come face-to-face with the unknown.

Here is a story that lands closer to home for me. One morning at 6:00am, in the summer of 2010, my wife arrived for work. The office building where she worked is just off US-131 on D Avenue, in Kalamazoo, Michigan. A co-worker had pulled into the parking lot there, as well; they both exited their vehicles and began walking toward the building. As they neared the entrance, they were startled by a long, high-pitched scream. My wife turned to her co-worker and asked, "WHAT was that!" The co-worker, with raised eyebrows answered, "I have no idea!" My wife then said, "Do you think it was a coyote?" Shaking his head, the co-worker who is a hunter, who also hunts coyotes, said, "No. I know the cries of coyotes and THAT was no coyote!"

Briefly, they surveyed the high brush that bordered the far side of the lot and after seeing nothing, continued into the building for their work day.

When she came home, she excitedly told me about the incident so I pulled up GOOGLE satellite maps of the surrounding properties, which showed the area was a combination of fields and woods with a stream running through it. There are a few homes in the area but they are sparsely located. There have been sightings of Bigfoot in, and around, the area.

Not satisfied with GOOGLE maps, I drove to the area which is only 10 minutes away. What I saw is what is typical of this county: wooded areas and fields with an occasional house. Bigfoot could easily move through the area remaining largely unseen. Today, however, more houses are appearing due to the ongoing urban sprawl that now infects smaller cities and towns.

The most recent sighting of a Bigfoot-like creature was seen by multiple witnesses: "There seems to be some big interest about what happened this morning off US-131 near D Ave in Kalamazoo. Apparently there was some kind of large, mysterious animal which caused drivers to stop their cars and bust out their phones according to one woman who posed the question on Facebook: "About 8:30 this morning I was driving down 131 almost to the D Ave exit going north when I seen maybe a half a dozen cars and 1 semi pulled over just barely on the off-ramp and everyone was out of their cars with their phones looking at an animal that just crossed the highway. I couldn't see it myself! Has anyone heard about this and know what it was?!" - Courtesy of WKFR 103.3

WKFR later offered an update on their site: "UPDATE AS OF 9/13/19: We may never know just exactly WHAT was seen or happened this past Monday morning. One man in our comment section said there was some kind of accident and a car was submerged in a swamp, but I don't know why people would get out of their cars and record something like that. A lot of people still claim to have seen something Bigfoot related in that area. We STILL have no clue what people recorded that morning because nobody has sent us anything. Until we do get some info, we may never know, and this tale will become legend."

I have not seen any pictures, or videos, posted from that incident. In this age of instant gratification via the internet and its various social sites, I find it odd that there was not a flurry of people posting their pictures or videos of the sighting. After the update WKFR posted on their site, I contacted them to see if anyone had sent them pictures, or other content since then and their response was, "there has been nothing further."

The Bigfoot Field Researchers Organization (BFRO) lists sightings that were reported near the area:

May 2002 (Class A) - Nighttime sighting by four witnesses, near Plainwell

Spring 1989 (Class A) - Multiple encounters, multiple witnesses, dating back to 60's

Spring 1969-1979 (Class A) - Multiple childhood interactions recalled by a woman who grew up near Kalamazoo

Michigan has approximately 36 million acres of land, with about 20 million acres classified as forested, covering around 53% of the state's total land area. This forested area has increased by over a million acres since 1980.

The Upper Peninsula, while accounting for 29% of Michigan's land area, contains 45% of the state's forests. It. is a very low populated expanse of forests with just a little over 300,000 people living in a bit over 16,000 square miles of land, or 18 people per square mile, but 24% of that population live in the larger tourist areas.

Bigfoot certainly has enough room to easily roam throughout Michigan.

Chapter 7

While Bigfoot is most commonly associated with the forests of North America, stories of large, hairy, human-like creatures are not unique to this continent. Across the world, cultures from vastly different regions and backgrounds share tales of mysterious beings that resemble the creature known as Bigfoot. From the Yeti in the Himalayas to the Yowie in Australia, these legends of "wild men" appear in nearly every part of the globe, suggesting that the idea of an elusive, bipedal creature is a common thread in human folklore and history.

One of the most famous international counterparts to Bigfoot is the Yeti, also known as the "Abominable Snowman." The Yeti is said to inhabit the high mountain ranges of the Himalayas, particularly in Nepal, Bhutan, and Tibet. Described as a large, ape-like creature covered in white or reddish-brown fur, the Yeti has been part of Himalayan folklore for centuries. Local Sherpa communities believe the Yeti to be a spiritual being, both feared and revered for its strength and mysterious nature.

Western interest in the Yeti grew in the early 20th century when explorers and climbers began venturing into the Himalayas. One of the most famous Yeti-related expeditions took place in 1951 when British mountaineer Eric Shipton discovered a series of large footprints while climbing Mount Everest. Shipton's photographs of the prints, measuring about 13 inches long, sparked international intrigue, and soon, the Yeti became a popular subject in cryptozoological circles. Despite numerous expeditions over the years, no definitive evidence of the Yeti's existence has been found, but the stories persist, and the creature remains an enduring figure in Himalayan folklore.

Moving to the jungles of Southeast Asia, we encounter the Orang Pendek, a small, bipedal creature said to inhabit the forests of Sumatra, Indonesia. Unlike Bigfoot or the Yeti, the Orang Pendek is often described as much shorter—typically around three to four feet tall—but it shares many similarities with its larger counterparts, such as its human-like posture, ape-like appearance, and elusiveness. Reports of the Orang Pendek date back to the early 20th century, when Dutch colonists in Sumatra first heard stories of the creature from local tribes.

Sightings of the Orang Pendek have continued into modern times, with many reports coming from farmers, hunters, and explorers. Witnesses describe the creature as being covered in dark or reddish-brown hair, with a muscular build Some cryptozoologists believe that the Orang Pendek could be a surviving species of "Homo floresiensis," a small hominid species that lived on the nearby island of Flores and is believed to have gone extinct around 50,000 years ago. While no physical evidence of the Orang Pendek has been found, its continued presence in local folklore and eyewitness reports makes it one of the most compelling cryptids in Southeast Asia.

Australia, too, has its own version of Bigfoot, known as the Yowie. Descriptions of the Yowie vary, but it is generally portrayed as a large, hairy creature, standing between six and ten feet tall, with ape-like features and a powerful build. The Yowie is said to inhabit the remote wilderness areas of Australia, particularly in the eastern states of New South Wales and Queensland. Stories of the Yowie have been part of Aboriginal oral traditions for centuries, and like Bigfoot, the creature is often seen as a guardian of the natural world, with the ability to disappear into the wilderness at will.

European settlers in Australia began reporting encounters with the Yowie in the 19th century, and sightings have continued into the modern era. In recent years, numerous individuals have come forward with stories of seeing or hearing the Yowie while camping or hiking in the Australian bush. While skeptics argue that these encounters could be explained by known animals, such as kangaroos or large feral pigs, believers maintain that the Yowie is a distinct, unknown species that has yet to be documented by science.

In Russia and other parts of Eastern Europe, tales of the Almasty, a wild man said to roam the Caucasus Mountains, have been passed down for generations. The Almasty is often described as a large, human-like creature covered in dark hair, but unlike Bigfoot or the Yeti, the Almasty is frequently portrayed as being more human than ape-like in appearance. Some accounts even suggest that the Almasty can speak in a rudimentary language, adding to the mystery of its origins.

Reports of the Almasty gained attention in the Soviet Union during the mid-20th century, when researchers and explorers began investigating sightings in remote areas of the Caucasus and Siberia. Some researchers, such as Boris Porshnev, believed that the Almasty could be a surviving population of Neanderthals, living in isolation from modern humans. While this theory has not gained widespread acceptance, the idea that a relic population of ancient hominids could still exist in remote parts of the world is a tantalizing one for cryptozoologists.

Even in Africa, stories of mysterious, ape-like creatures persist. In the Congo Basin, there are tales of the Kikomba, a large, bipedal creature said to inhabit the dense rainforests. The Kikomba is described as a nocturnal, elusive being, similar in many ways to Bigfoot. Local tribes in the region believe that the Kikomba is a spirit of the forest, capable of vanishing into the trees when threatened. While there is little physical evidence to support the existence of the Kikomba, the stories have continued for generations, passed down through oral traditions.

One of the more intriguing aspects of these global legends is the consistency of certain details. Whether it's the Yeti in the Himalayas, the Orang Pendek in Sumatra, or the Yowie in Australia, these creatures are often described as large, bipedal, and covered in hair. They are elusive, intelligent, and often regarded as guardians of the natural world. These similarities raise interesting questions: Are these legends simply a product of human imagination, or could they be based on real encounters with unknown species?

Some anthropologists suggest that the global prevalence of "wild man" legends could be rooted in humanity's collective memory of coexisting with other hominid species. For much of human history, Homo sapiens were not the only human-like species on the planet. Neanderthals, Denisovans, and other hominid species once roamed the Earth, and it's possible that stories of these ancient encounters have survived in the form of folklore.

Others argue that these legends could be a way for cultures to express their relationship with the natural world. The wild man represents the untamed, primal forces of nature—something that cannot be controlled or fully understood. In many cultures, these creatures are not seen as threats, but rather as symbols of the wilderness itself, embodying the idea that there are still places on Earth that remain beyond the reach of civilization.

Regardless of their origins, the global prevalence of Bigfoot-like creatures adds a fascinating dimension to the mystery. While the search for Bigfoot in North America continues, researchers around the world are investigating their own local legends, hoping to uncover the truth behind these enduring stories.

Whether the answer lies in ancient hominids, undiscovered species, or the human psyche, one thing is clear: the wild man is a universal figure, a reminder that the world is still full of mysteries waiting to be explored, and a powerful presence despite its smaller size.

Chapter 8

As with many mysteries that captivate the public imagination, the Bigfoot phenomenon has not been immune to hoaxes, fabrications, and misunderstandings. From intentionally planted footprints to costumes and staged sightings, there have been numerous attempts to fake evidence of Bigfoot.

These hoaxes, whether for profit, fame, or sheer mischief, have complicated the search for the truth, making it difficult for researchers to separate genuine evidence from fabricated claims. In this chapter, we explore some of the most infamous Bigfoot hoaxes and how they've impacted the broader investigation into the creature's existence.

One of the earliest and most influential Bigfoot hoaxes occurred in 1958, when large footprints were discovered near a logging site in Bluff Creek, California. The prints, measuring an astonishing 16 inches in length, were initially heralded as groundbreaking evidence of a massive, unknown creature living in the area. The local media quickly dubbed the creature "Bigfoot," and the legend was cemented in the public consciousness. However, years later, it was revealed that the footprints were a hoax orchestrated by a man named Ray Wallace, whose family admitted after his death that he had created the prints using wooden cutouts.

Despite the Bluff Creek hoax, the incident had a lasting impact on the Bigfoot phenomenon. The name "Bigfoot" stuck, and the story inspired a wave of interest in the creature, with more sightings and footprint discoveries reported in the years that followed. For many, the hoax served as a cautionary tale, highlighting the need for skepticism and careful investigation in the search for Bigfoot. Yet, the excitement generated by the

Bluff Creek incident also fueled a broader cultural fascination with the idea of an elusive, bipedal creature hiding in the wilderness.

Another notorious hoax came in 2008, when two men from Georgia, Rick Dyer and Matthew Whitton, claimed to have found the body of a Bigfoot. The pair released photographs of what appeared to be a large, hairy creature frozen in ice, sparking a media frenzy. They held a press conference, announcing that they had discovered the remains of the legendary creature in the woods of northern Georgia and that DNA tests were underway to confirm its authenticity.

For a brief moment, it seemed as though definitive proof of Bigfoot had finally been found. But as the story gained traction, skepticism grew. When the so-called Bigfoot body was finally thawed for examination, it was quickly revealed to be a costume stuffed with animal parts and frozen in ice. The hoax unraveled, and Dyer and Whitton admitted that the entire thing had been a fabrication. While the stunt briefly brought them fame, it also brought scorn from the Bigfoot community and the general public.

Dyer, however, wasn't finished with Bigfoot hoaxes. In 2014, he resurfaced, once again claiming to have shot and killed a Bigfoot in Texas. This time, he toured with the supposed body, charging people to see it. As with the 2008 hoax, it was eventually revealed that the body was another fake, this time made of latex. Dyer's repeated attempts to fool the public contributed to widespread skepticism about Bigfoot sightings and evidence, making it even more difficult for legitimate researchers to be taken seriously.

Misidentifications of known animals have also contributed to confusion in the Bigfoot search. One common example is the black bear, which, when standing on its hind legs, can resemble a large, upright creature. In regions where bears are prevalent, it's easy to imagine how a glimpse of a bear moving through the trees could be mistaken for something more mysterious, especially if seen from a distance or in poor lighting conditions.

In one famous case from 1976, a Bigfoot sighting in the Gifford Pinchot National Forest in Washington was reported by a group of campers who described seeing a large, dark figure moving through the trees. When investigators arrived on the scene, they found tracks that led them to a black bear den. It appeared that the bear, moving upright, had been the source of the sighting. Such incidents are common, with bears, large elk, and even humans in heavy clothing all contributing to misidentified Bigfoot sightings over the years.

In addition to animals, human-made objects or structures in the wilderness can also lead to misidentifications. In some cases, people have mistaken large tree stumps or shadows for Bigfoot, especially in the eerie twilight hours when visibility is low. Some have even reported strange sounds, such as wood-knocking or howling, only to discover later that the noises were caused by tree branches rubbing together or other natural phenomena.

While hoaxes and misidentifications have certainly muddied the waters of Bigfoot research, they've also served as valuable lessons for investigators. The need for thorough documentation, skepticism, and scientific rigor has become paramount in the search for the creature. Modern Bigfoot researchers are often the first to debunk hoaxes or misidentifications, recognizing that credible evidence is the only path to being taken seriously by the wider scientific community.

Despite the setbacks caused by hoaxes, legitimate Bigfoot enthusiasts and researchers continue to pursue their investigations with passion and dedication. Advances in technology, such as trail cameras, drones, and environmental DNA testing, have given modern researchers tools that previous generations could only dream of. With these advancements, there is hope that one day, conclusive evidence of Bigfoot will be found.

But the legacy of hoaxes remains a cautionary tale. For every legitimate sighting or piece of evidence, there is the risk of another hoax that could undermine the credibility of the entire field. In the world of cryptozoology, where belief and skepticism are often in tension, the importance of maintaining scientific integrity cannot be overstated.

As we look ahead to the future of Bigfoot research, it's clear that those who seek fame, fortune, or simply a good laugh will continue to blur the line between myth and reality. But for those who truly believe in the possibility of discovering an unknown species, the search will go on, fueled by curiosity, hope, and the occasional moment of genuine mystery.

Chapter 9

As the search for Bigfoot has evolved, so too have the methods and tools used by researchers. In the early days of Bigfoot investigations, sightings were largely anecdotal, based on eyewitness accounts and footprint casts. Today, however, advances in technology have given modern researchers new ways to track, document, and analyze potential evidence of the elusive creature. From high-tech cameras to DNA analysis, these tools have brought a new level of sophistication to the hunt for Bigfoot, even as the creature itself remains as elusive as ever.

One of the most widely used tools in modern Bigfoot research is the trail camera, or game camera. These motion-activated devices are commonly used by hunters to monitor wildlife, and they have become an essential tool for Bigfoot investigators. Trail cameras are set up in remote areas where sightings have been reported, allowing researchers to capture images or video footage of any animals—or potential Bigfoot—that pass by.

Because the cameras operate passively and can remain in place for long periods, they provide an opportunity to capture evidence without the need for human presence, which might otherwise scare off a shy or cautious creature.

While trail cameras have successfully captured images of wildlife—bear, deer, mountain lions, and more—there has yet to be a definitive image of Bigfoot captured in this way. Some researchers believe that the creature might be intelligent enough to avoid cameras, perhaps sensing the infrared light emitted by the devices or recognizing the unnatural placement of a camera in the wilderness. Others argue that the lack of clear footage could simply be due to the creature's extreme rarity or its ability to blend into its surroundings. Nevertheless, trail cameras remain a staple in the Bigfoot research toolkit, and investigators continue to set them up in remote areas in the hopes of one day capturing that elusive footage.

In addition to trail cameras, drones have become an increasingly valuable tool in the search for Bigfoot. Equipped with high-definition cameras and thermal imaging technology, drones allow researchers to cover vast areas of wilderness quickly and efficiently. Drones can fly over rugged, inaccessible terrain that would be difficult or impossible for humans to traverse on foot, giving researchers a bird's-eye view of potential hotspots for Bigfoot activity.

Thermal imaging, in particular, has been a game-changer for Bigfoot researchers. By detecting heat signatures, thermal cameras can identify animals—or potentially Bigfoot—that are otherwise hidden by darkness or dense foliage. In recent years, several researchers have reported capturing thermal images of large, bipedal figures in remote areas, although these images are often inconclusive and difficult to verify. Still, thermal imaging remains one of the most promising technologies for tracking and identifying elusive creatures in the wild.

One of the most exciting developments in the hunt for Bigfoot is the use of eDNA sampling. This relatively new technique allows scientists to collect DNA from the environment, such as soil, water, or air, without the need for direct contact with the creature itself. Animals leave traces of their DNA in their surroundings through skin cells, hair, feces, or other biological material, and eDNA can be used to identify the species that are present in a given area.

eDNA has already proven to be a powerful tool in the study of aquatic environments, where it has been used to detect the presence of rare or elusive species such as the great white shark and the endangered European eel. In the context of Bigfoot research, eDNA offers the potential to detect the creature's presence in areas where it has been sighted, without the need for a body or other physical remains. While no definitive Bigfoot DNA has been identified through eDNA sampling so far, researchers are optimistic that this technology could eventually provide the evidence needed to confirm the creature's existence.

Another technological advancement that has had a significant impact on Bigfoot research is digital audio recording. As mentioned in earlier chapters, many Bigfoot sightings are accompanied by strange vocalizations—howls, screams, wood knocks, and other sounds that do not match the calls of known animals. By placing high-quality digital audio recorders in remote areas, researchers can capture these sounds and analyze them for patterns or signatures that might suggest the presence of an unknown species.

Audio experts have analyzed several well-known recordings, such as the Sierra Sounds discussed in previous chapters. Some suggest that the vocalizations exhibit characteristics not found in human speech or animal calls. While these recordings are intriguing, they remain open to interpretation, and more data is needed to draw definitive conclusions. Nevertheless, audio recording remains a valuable tool for researchers hoping to document the presence of Bigfoot in the wilderness.

In addition to capturing images, sounds, and DNA, modern Bigfoot researchers have also begun to use geographic information systems (GIS) and data analysis to track and map sightings. By compiling reports from witnesses and overlaying them on maps, researchers can identify patterns in Bigfoot sightings, such as specific regions or habitats where the creature is more likely to be encountered. GIS technology allows researchers to visualize these patterns in ways that were not possible before, offering new insights into the creature's potential range and behavior.

One notable example of this approach is the BFRO, which maintains a comprehensive database of Bigfoot sightings across North America. By analyzing the data, the BFRO has identified certain areas, such as the Pacific Northwest and parts of the Appalachian Mountains, as hotspots for sightings, suggesting that these regions might be home to a larger population of the creatures. While the database relies on eyewitness reports, which are inherently subjective, it provides a valuable resource for researchers trying to identify trends and focus their investigations.

While modern technology has undoubtedly advanced the search for Bigfoot, it has also raised new challenges. For example, the widespread availability of digital editing software has made it easier than ever for hoaxers to create fake Bigfoot footage or audio recordings. This has made it more difficult for legitimate researchers to distinguish between genuine evidence and fabrications, further complicating the already controversial field of Bigfoot research.

Moreover, even with all the advances in technology, the fundamental challenge remains the same: Bigfoot, if it exists, is an incredibly elusive creature. It inhabits some of the most remote and inaccessible areas on Earth, where human presence is rare and fleeting. Despite the best efforts of researchers armed with the latest technology, Bigfoot continues to evade capture, both physically and in terms of conclusive evidence.

However, for those dedicated to the search, the absence of proof is not proof of absence. The tools of modern science offer new opportunities for discovery, and the hope of one day finding conclusive evidence of Bigfoot remains strong. As technology continues to improve, researchers believe it's only a matter of time before we capture the definitive image, sound, or DNA sample that will finally solve the mystery.

Chapter 10

Bigfoot's journey from obscure folklore to global pop culture icon is a testament to the creature's enduring appeal. What began as regional tales passed down through generations has evolved into a worldwide phenomenon, with Bigfoot taking on a life of its own in movies, television, books, and even marketing campaigns. In this chapter, we explore how Bigfoot has been portrayed in popular culture and how the creature has become a symbol of mystery, adventure, and the unknown.

The rise of Bigfoot as a cultural icon can be traced back to the mid-20th century, when the modern Bigfoot phenomenon first gained traction. As discussed in previous chapters, the discovery of giant footprints in Bluff Creek, California, in 1958, and the subsequent release of the Patterson-Gimlin film in 1967 helped thrust

Bigfoot into the national spotlight. News outlets covered the story, and soon the creature was a household name. This marked the beginning of Bigfoot's transformation from regional cryptid to pop culture staple.

One of the earliest portrayals of Bigfoot in film came in 1972 with "The Legend of Boggy Creek," a docudrama that mixed real-life interviews with dramatic reenactments of Bigfoot sightings in Fouke, Arkansas. The film was a low-budget production, but it became a cult hit, capitalizing on the public's growing fascination with Bigfoot. "The Legend of Boggy Creek" depicted the creature as a mysterious and fearsome being, lurking in the swamps and terrorizing small-town residents. The film's success paved the way for a wave of Bigfoot-related movies throughout the 1970s and 1980s.

One of the most well-known Bigfoot films from this era is "Sasquatch: The Legend of Bigfoot" (1977), a semi-documentary film that follows a group of explorers searching for the elusive creature in the Pacific Northwest. The film, like many others of its time, portrayed Bigfoot as a dangerous, almost mythical figure—something to be feared rather than admired. This depiction of Bigfoot as a menacing, wild creature became a common theme in films of the period, reinforcing the idea that Bigfoot was a terrifying denizen of the wilderness.

However, not all portrayals of Bigfoot in pop culture have been menacing. In the 1980s, Bigfoot underwent something of an image makeover, thanks in large part to the film "Harry and the Hendersons" (1987). This family-friendly comedy portrayed Bigfoot, or "Harry," as a gentle giant who befriends a suburban family. The film humanized Bigfoot, making him less of a cryptid and more of a misunderstood creature who just wanted to live in peace. "Harry and the Hendersons" was a commercial success and introduced Bigfoot to a new generation of children and families, transforming the creature into a more lovable and sympathetic figure.

Following the success of "Harry and the Hendersons," Bigfoot began to appear in a wider range of media, from cartoons to commercials. In the 1990s, Bigfoot became the mascot for Jack Link's Beef Jerky, starring in a long-running series of humorous commercials where the creature, now referred to as "Sasquatch," was the hapless subject of pranks by humans. These lighthearted portrayals helped solidify Bigfoot's place in pop culture as a figure of both mystery and humor—and the beef jerky wasn't bad either.

The 2000s saw a resurgence of interest in Bigfoot, fueled by reality television and paranormal investigation shows. Programs like "Finding Bigfoot" on Animal Planet and "MonsterQuest" on the History Channel brought Bigfoot back into the spotlight, offering viewers a mix of eyewitness accounts, scientific investigations, and dramatic reenactments of sightings. While these shows often left audiences with more questions than answers, they succeeded in keeping the Bigfoot legend alive and sparking new interest in the creature.

Bigfoot's influence has extended beyond film and television into the world of literature, where the creature has been featured in everything from horror novels to children's books. In cryptozoology-focused books, Bigfoot is often presented as a subject of serious investigation, with authors detailing the creature's possible origins, behavior, and sightings. Meanwhile, fiction writers have taken creative liberties with the Bigfoot legend, imagining the creature in all sorts of scenarios—from horror to romance.

In recent years, Bigfoot has even become a part of internet culture, with memes, viral videos, and fan theories spreading across social media platforms. The creature's enigmatic nature makes it a perfect subject for internet humor, and the image of Bigfoot—whether it's a blurry figure in the woods or a friendly giant—has become a symbol of the unknown in a digital age where information is readily available. The idea that something as elusive as Bigfoot could still remain hidden in the modern world adds to the creature's mystique, and online communities dedicated to Bigfoot continue to thrive.

Bigfoot's influence in pop culture also extends into the world of tourism and commerce. In regions where Bigfoot sightings have been reported, towns have embraced the creature as a local icon, using it to attract tourists and boost the local economy. Places like Willow Creek, California (near Bluff Creek, where the famous footprints were found), and Fouke, Arkansas (home of the Boggy Creek Monster), have turned Bigfoot into a tourist attraction, hosting festivals, museums, and guided tours. Bigfoot-themed merchandise—from T-shirts to coffee mugs—is readily available, and Bigfoot-themed businesses, such as breweries and restaurants, have popped up across the country.

The commercial appeal of Bigfoot is perhaps best illustrated by the success of Bigfoot-themed reality TV shows, documentaries, and merchandise. For many people, Bigfoot has become more than just a cryptid—it's a brand. The creature's image can be found on everything from socks to bumper stickers, and it has become a symbol of adventure, the wilderness, and the thrill of the unknown.

But why does Bigfoot resonate so deeply with people across cultures and generations? Part of the creature's appeal lies in its mystery. Bigfoot represents the idea that there are still places on Earth that remain untamed, where the rules of the modern world don't apply. In a time when satellite images can map even the most remote corners of the globe, the possibility that a large, intelligent creature could still be out there, undiscovered, is both thrilling and comforting. Bigfoot reminds us that the world is still full of surprises and that there are still frontiers left to explore.

Additionally, Bigfoot taps into a deep-rooted human fascination with the wild and the unknown. The creature embodies the tension between civilizations and nature, between the familiar and the untamed. In Bigfoot, we see a reflection of our own primal past, a reminder that humanity, too, once roamed the forests as part of the natural world. For some, Bigfoot represents a connection to that ancient heritage—a living link to a time when humans and nature were more closely intertwined.

As Bigfoot continues to evolve as a cultural icon, its significance is likely to change with the times. Whether as a subject of scientific inquiry, a figure of folklore, or a commercial mascot, Bigfoot remains a symbol of the unknown—a creature that invites us to question what we think we know about the world. And as long as there are unexplored forests and unanswered questions, the legend of Bigfoot will endure, capturing the imagination of those who dare to wonder what might be lurking just beyond the edge of the known.

Chapter 11

The legend of the Sister Lakes Monster is a fascinating chapter in the history of Bigfoot lore. Nestled near Dowagiac, Michigan, the Sister Lakes region became a hotbed of excitement and fear in the 1960s, as numerous residents reported sightings of a large, hairy, bipedal creature roaming the area. For those who lived nearby, it was more than just a story—it was a real, tangible mystery that hit close to home.

For my wife and her family, the story of the Sister Lakes Monster holds a deeply personal connection. Her grandfather, living in Allegan, Michigan, at the time, was captivated by the unfolding drama. The tales of a hulking, shadowy figure roaming the wooded areas and scaring locals piqued his curiosity, leading to an impulsive late-night adventure. He packed his family—including my wife, her mother, and brother—into the car and drove down to the area to see if they could catch a glimpse of the elusive creature.

As my wife told the story, I couldn't help but laugh at her vivid recollection of that tense and thrilling night. "Did you look around?" I asked, imagining them bravely scanning the dark woods for signs of the monster. Her response was immediate and emphatic: "Hell, no, we didn't get out of the car!" The thought of an 11- or 12-year-old, sitting wide-eyed in the safety of the vehicle, was both humorous and relatable. Even though the adventure didn't yield a sighting, the tension in the air and the knowledge that something could be out there made a lasting impression.

The Sister Lakes Monster became the stuff of local legend during the summer of 1964. Reports described it as standing over seven feet tall, covered in dark hair, and possessing an unsettling, humanlike gait.

Some witnesses claimed it had glowing eyes that pierced the darkness, adding an eerie element to the sightings. It was said to prowl around rural homes, peer through windows, and sometimes emit terrifying, guttural sounds.

The creature gained widespread attention when local law enforcement became involved. Police patrolled the area, responding to calls from frightened residents and investigating the strange tracks and other signs left behind. The media quickly picked up on the story, drawing cryptid enthusiasts and curious thrill-seekers to the quiet community of Sister Lakes.

Despite the fear and intrigue, no conclusive evidence of the creature was ever found. Theories abounded: some suggested it was a misidentified bear, while others believed it to be an elaborate hoax. For the locals, however, it didn't matter whether the monster was real or imagined—it left a permanent mark on their collective memory.

The Sister Lakes Monster is one of many regional Bigfoot stories that highlight the universal appeal of such legends. Like my wife's family, countless others were drawn to the mystery, whether through late-night drives, community conversations, or simply peering out of their windows and wondering what might be lurking in the dark.

For me, this story illustrates the personal and familial bonds these legends can create. My wife's retelling of her childhood experience brings a sense of humor and humanity to the broader mystery. It reminds us that while cryptids like the Sister Lakes Monster might remain elusive, the stories they inspire have a way of bringing people together, fueling curiosity, and creating lasting memories—even if you never leave the safety of the car.

Chapter 12

The search for Bigfoot is about more than solving a mystery. It's also a call to protect the wild spaces that have long inspired curiosity and awe. The dense forests, rugged mountains, and remote swamps where the creature is believed to reside are among the most biodiverse and fragile ecosystems on the planet. While cryptozoologists focus on uncovering evidence of an elusive figure, their work underscores a broader and more pressing concern: the preservation of these vital habitats. Bigfoot, in a sense, has become a symbol of these untamed spaces, a representation of the interconnectedness between humans and nature.

While films like *Harry and the Hendersons* have humanized Bigfoot for a broad audience, the creature remains, for many, a symbol of untamed wilderness and the need to protect it. Whether it's the deep woods of the

Pacific Northwest, the isolated peaks of Appalachia, or the shadowed valleys of the Rocky Mountains, these habitats tell their own stories. These landscapes are home to countless creatures—some documented and others still unknown. Protecting these areas isn't just about safeguarding the possibility of Bigfoot's existence; it's about ensuring the survival of diverse ecosystems and the wildlife that thrives within them.

Even as Bigfoot remains elusive, scientists continue to discover new species in places once thought to be fully explored. From rare amphibians to previously undocumented plants, these findings remind us that there is still much left to learn about the natural world. The search for Bigfoot, though often seen as fringe science, has occasionally played a surprising role in conservation efforts.

Researchers drawn to Bigfoot hotspots frequently uncover valuable data unrelated to the creature itself. Expeditions often document wildlife migration patterns, map underexplored regions, and even stumble upon species thought extinct.

For instance, efforts to find the Orang Pendek in Sumatra have yielded significant insights into the biodiversity of the region's rainforests. Similarly, the search for Bigfoot in the Pacific Northwest has sparked broader conversations about deforestation and its impact on the environment. What began as a quest for one enigmatic creature often evolves into a deeper understanding of the ecosystems that support countless lives.

Bigfoot's legendary ability to evade detection may be rooted in the dense, wild habitats it's said to inhabit. These same habitats, under threat from deforestation, urban expansion, and climate change, are also some of the most vital to the planet's health. Forests act as the lungs of the Earth, mountain watersheds sustain human and animal life, and untouched wildernesses provide sanctuary for species that might not survive elsewhere. Bigfoot has become a stand-in for these environments, a figure that represents the delicate balance of nature and the consequences of its disruption.

For many who believe in the creature, Bigfoot is not just a cryptid but a guardian of the wild. This belief often leads to advocacy, with enthusiasts championing efforts to protect forests, wetlands, and other spaces they see as potential homes for the creature. By framing Bigfoot as a living testament to the wonders of wilderness, researchers and believers alike inspire conversations about conservation, urging communities to reconsider how they interact with the natural world.

The legend of Bigfoot challenges us to rethink what we know about nature and our place within it. It's a reminder that the Earth still holds mysteries.

Chapter 13

Bigfoot's elusiveness isn't merely a matter of chance; it may reflect the creature's profound adaptation to its environment, showcasing a blend of intelligence, instinct, and physiological traits that allow it to evade detection. Many researchers speculate that its behavioral patterns—its movements, diet, and social structures—are finely tuned to the rugged, remote terrains it inhabits. Exploring these possibilities helps illuminate not only why Bigfoot remains undiscovered but also how it may interact with its surroundings.

One of the most compelling arguments for Bigfoot's ability to remain hidden is its presumed intelligence. If the creature is an advanced primate or a hominid relative, it likely possesses cognitive abilities on par with, or even surpassing, great apes. This intelligence would allow it to recognize and avoid potential threats, including humans. Stories of Bigfoot reportedly avoiding trail cameras or detecting researchers before they approach suggest a level of awareness not typically associated with ordinary wildlife. It might explain the frustrating lack of clear evidence despite decades of effort by investigators.

This presumed intelligence would likely extend to its understanding of its habitat. In areas of dense forest, Bigfoot could navigate with ease, moving silently through brush and using the terrain to obscure itself. Many researchers believe it employs natural cover such as fallen logs, thick undergrowth, and rocky outcroppings to conceal its movements. Some sightings describe the creature disappearing almost instantly into the wilderness, leaving observers baffled. Its agility and speed make even a fleeting glimpse a rare occurrence.

Bigfoot's diet also plays a role in its ability to thrive undetected. As an omnivore, it likely consumes a wide variety of foods, depending on what the local environment provides. Reports of foraged berries, uprooted plants, and scavenged animal remains suggest a creature well-adapted to the seasonal cycles of its habitat. Some evidence points to opportunistic feeding habits, such as raiding dumpsters or scavenging scraps left behind by humans, which might account for occasional sightings near populated areas. These behaviors align with the survival strategies of other intelligent animals, such as bears and raccoons, which take advantage of available resources while minimizing risk.

The social structure of Bigfoot, if it exists, is another intriguing factor. Many sightings describe single individuals, leading to speculation that the creature may be solitary by nature. However, accounts of multiple individuals—particularly juveniles—suggest that Bigfoot might have a loose social organization similar to that of orangutans or smaller primates. Solitary males might roam vast territories, while females and young remain in smaller, family-based groups. This behavior would reduce the likelihood of large gatherings, further complicating efforts to locate and study them.

Seasonal migration might also explain why Bigfoot sightings occur in waves. Researchers have noted that reports often coincide with changes in weather or food availability, suggesting that the creature might follow traditional wildlife patterns, moving between lowland and highland areas or along waterways. This mobility, combined with its ability to avoid detection, could explain why even the most dedicated investigators struggle to establish consistent evidence of its presence.

Another factor contributing to Bigfoot's secrecy is the vastness and inaccessibility of its potential habitat. Remote forests, mountainous regions, and swamps provide a level of isolation that limits human intrusion. Even in the age of satellite imagery and GPS, these areas remain underexplored. The Pacific Northwest, in particular, is home to some of the most rugged terrain in North America. Its towering trees, deep valleys, and thick undergrowth create a natural fortress where wildlife can flourish largely undisturbed. If Bigfoot resides in such places, it benefits from the same protections that have allowed countless other species to go unnoticed by human eyes.

The question of why no physical remains have been found is often raised by skeptics, but it's worth noting that nature is exceptionally efficient at reclaiming its own. In the wild, decomposition occurs rapidly, aided by scavengers, insects, and microorganisms. Bones are often scattered and absorbed into the environment over time. Coupled with the possibility that Bigfoot buries its dead—either for hygienic purposes or as part of a social ritual—this would make finding remains exceedingly rare. Even for known species, such as bears and mountain lions, encountering intact skeletons is an uncommon event, despite their substantial populations.

One final consideration is the possibility of Bigfoot's deliberate avoidance of humans. Many Indigenous traditions describe the creature as shy, elusive, and wary of human contact. This behavior aligns with the survival instincts of other large mammals, which generally avoid confrontations unless threatened. In some accounts, Bigfoot is said to observe humans from a distance, retreating as soon as it senses attention. This combination of curiosity and caution could reflect a creature that understands the dangers posed by humans and has adapted to live in the shadows of human activity.

Whether these traits are enough to explain Bigfoot's enduring mystery remains to be seen, but they offer a compelling framework for understanding its potential behavior and ecology.

The creature's elusiveness is not just a matter of luck or coincidence; it may be the result of millions of years of evolution, shaping an animal perfectly adapted to survive in the world's most remote and challenging environments. For researchers, this presents both a challenge and an opportunity: the need to outthink a creature that, if real, has mastered the art of disappearing. The search for Bigfoot is not just about finding a cryptid; it's about uncovering the intricate relationships between life, landscape, and survival in the natural world.

Chapter 14

As cryptozoology has evolved over the years, so too has the role of technology in shaping both the pursuit of and belief in creatures like Bigfoot. From early print media to modern digital platforms, technology has played a critical role in spreading the Bigfoot legend, as well as in driving the search for evidence. In this chapter, we'll explore how technology, from cameras to the internet, has influenced the way people think about Bigfoot and how it has shaped both the evidence-gathering process and public perception of the creature.

In the early days of the Bigfoot phenomenon, technology was limited to traditional forms of media such as newspapers, radio, and television. These early media outlets played a key role in popularizing the legend of Bigfoot, spreading stories of sightings, footprints, and encounters to a wider audience. For example, the discovery of large footprints in Bluff Creek, California, in 1958 was widely reported in local newspapers, and the term "Bigfoot" was coined by journalists covering the story. This coverage helped bring the creature into the national spotlight, and soon Bigfoot was a household name.

However, it was the release of the Patterson-Gimlin film in 1967 that truly cemented Bigfoot's place in popular culture. Captured on 16mm film, the footage of a large, hairy creature walking through the forests of Northern California quickly became one of the most iconic and controversial pieces of evidence in the history of cryptozoology. The grainy, shaky footage was broadcast on television and analyzed by experts and enthusiasts alike, with many believing it to be the most compelling proof of Bigfoot's existence.

The impact of the Patterson-Gimlin film highlights the power of visual media in shaping belief. Prior to the film, Bigfoot sightings were largely anecdotal, stories passed down through word of mouth or written in newspaper articles. The film, however, provided a visual representation of the creature that made the legend feel more tangible and real to many viewers. Even though the authenticity of the footage has been debated for decades, it remains a powerful symbol of the search for Bigfoot, showing how technology can bring cryptid sightings to life in a way that words alone cannot.

As technology advanced, so too did the methods used to search for Bigfoot. The development of trail cameras, for example, revolutionized wildlife observation, allowing researchers to capture images of animals in their natural habitats without human interference. These motion-activated cameras have become a key tool in the search for Bigfoot, offering the possibility of capturing definitive photographic evidence of the creature. However, despite the widespread use of trail cameras in Bigfoot research, no conclusive images of the creature have been captured to date, leading some researchers to speculate that Bigfoot may be adept at avoiding detection by these devices.

Thermal imaging technology has also played a significant role in modern Bigfoot research. Thermal cameras, which detect heat signatures, have been used by researchers to scan large areas of forest in search of unknown animals. These devices allow researchers to track movement in the dark or through dense foliage, providing a way to detect creatures that might otherwise remain hidden. While thermal imaging has captured intriguing heat signatures that some researchers believe could belong to Bigfoot, no definitive conclusions have been reached, as it is difficult to determine the source of the heat signatures without further evidence.

One of the most profound changes in the search for Bigfoot has been the rise of the internet and social media. The internet has democratized access to information and allowed people from all over the world to share their experiences, theories, and evidence related to Bigfoot. Online forums, websites, and social media platforms have become hubs for cryptozoologists, amateur researchers, and enthusiasts to discuss sightings, share photos and videos, and organize expeditions. This has created a global community of Bigfoot believers, allowing them to connect and collaborate in ways that were not possible before the digital age.

Social media, in particular, has had a significant impact on the spread of Bigfoot-related content. Platforms like YouTube, Instagram, and TikTok have become popular outlets for sharing footage of potential sightings, with videos of blurry figures, strange sounds, and unexplained footprints going viral and reaching millions of viewers. While much of this content is quickly debunked or dismissed as hoaxes, the viral nature of social media has helped keep Bigfoot in the public eye and has introduced a new generation to the legend.

However, the rise of digital technology has also presented new challenges for Bigfoot researchers. The widespread availability of photo and video editing software has made it easier than ever for hoaxers to create convincing fake evidence, from doctored photographs to digitally manipulated videos. This has made it increasingly difficult to distinguish genuine sightings from fabrications, and it has led to a growing sense of skepticism within the Bigfoot community.

For example, several viral videos that initially appeared to show compelling evidence of Bigfoot were later revealed to be hoaxes. In one case, a viral video purportedly showing a large, dark figure moving through the trees was later discovered to be a man in a gorilla costume. In another case, a photograph of what appeared to be a Bigfoot in the snow was revealed to be a creative use of perspective, with the "creature" actually being a small statue placed in the foreground. These incidents have contributed to a general distrust of digital evidence, as it has become increasingly difficult to verify the authenticity of photos and videos in the age of digital manipulation.

At the same time, technology has opened up new avenues for scientific investigation into Bigfoot. DNA analysis, for example, has become a key tool in wildlife research, allowing scientists to identify species through genetic material such as hair, skin, or feces. eDNA, in particular, has shown promise in detecting rare or elusive species by analyzing traces of DNA left behind in the environment. While no definitive Bigfoot DNA has been identified, researchers continue to explore the potential of genetic analysis as a way to verify the creature's existence.

The internet has also facilitated the development of databases and mapping tools that allow researchers to track Bigfoot sightings over time. Websites like the BFRO maintain extensive databases of reported sightings, allowing users to search for reports by location, date, or specific details. This data can be used to identify patterns in sightings, such as areas with high concentrations of reports or specific environmental conditions that may be linked to Bigfoot activity. GIS and mapping software have enabled researchers to visualize these patterns in new ways, providing valuable insights into the creature's possible range and behavior.

Despite the challenges posed by technology, the search for Bigfoot continues to be shaped by the tools available to researchers. Whether through trail cameras, thermal imaging, DNA analysis, or digital databases, technology has both expanded the possibilities for discovery and complicated the process of verifying evidence. For believers, the hope remains that technology will one day provide the definitive proof needed to confirm Bigfoot's existence. For skeptics, the increasing sophistication of hoaxes and digital manipulation reinforces the need for rigorous scrutiny and scientific investigation.

Ultimately, technology has both amplified the mystery of Bigfoot and brought it closer to resolution. As researchers continue to refine their methods and new technologies emerge, the search for Bigfoot will undoubtedly evolve, offering new opportunities for discovery and new challenges to overcome. Whether or not the creature is ever found, the role of technology in shaping belief and evidence will remain a central part of the Bigfoot story.

Chapter 15

While the legend of Bigfoot is most closely associated with North America, stories of large, hairy, human-like creatures are surprisingly common across the globe. From the Yeti of the Himalayas to the Orang Pendek of Indonesia, nearly every culture has its own version of the mysterious wild man—a creature that lives on the fringes of human society, evading detection while sparking curiosity and fear. In this chapter, we'll explore some of the most famous international counterparts to Bigfoot and examine how these creatures fit into the broader context of global folklore and belief.

Perhaps the most famous of Bigfoot's global cousins is the Yeti, also known as the "Abominable Snowman." The Yeti is said to inhabit the high mountains of the Himalayas, particularly in Nepal, Tibet, and Bhutan. Descriptions of the Yeti vary, with some accounts depicting it as a towering, ape-like figure covered in white or reddish-brown fur, while others describe it as smaller and more human-like in appearance. What all stories agree on, however, is the creature's mysterious and elusive nature, as well as its connection to the remote, snow-covered peaks of the world's tallest mountain range.

The Yeti has been a part of Himalayan folklore for centuries, with indigenous Sherpa and Tibetan communities viewing it as a powerful, sometimes spiritual figure. In local mythology, the Yeti is often portrayed as a guardian of the mountains, a being that both protects and warns against trespassing into sacred or dangerous areas. Unlike the menacing depictions of Bigfoot in some North American stories, the Yeti is often seen as a more neutral, or even protective, presence, though encounters with the creature are still said to invoke fear.

Western interest in the Yeti dates back to the early 20th century, when European mountaineers began exploring the Himalayas. In 1921, during a British expedition to Mount Everest, the leader of the team, Lieutenant Colonel Charles Howard-Bury, reported seeing large, mysterious footprints in the snow. His Sherpa guides attributed the prints to the Yeti, sparking a media sensation. The term "Abominable Snowman" was later coined by a British journalist, and the Yeti quickly became one of the most famous cryptids in the world.

In the decades that followed, numerous expeditions were launched to search for the Yeti, with varying degrees of success. In 1951, British mountaineer Eric Shipton famously photographed a series of large footprints on the slopes of Mount Everest, further fueling the Yeti legend. While no definitive evidence of the creature has been found, the Yeti remains a central figure in Himalayan folklore, and sightings and footprints continue to be reported by climbers and locals alike.

Moving from the snowy peaks of the Himalayas to the tropical forests of Southeast Asia, we encounter the Orang Pendek, a small, bipedal creature said to inhabit the jungles of Sumatra, Indonesia. Unlike Bigfoot or the Yeti, the Orang Pendek is typically described as much shorter, standing about three to five feet tall, with a stocky, muscular build and dark hair covering its body. The Orang Pendek is often described as being more human-like than ape-like, with an upright posture and a sense of intelligence or awareness in its eyes.

Reports of the Orang Pendek date back to the colonial period, when Dutch settlers in Sumatra heard stories of the creature from local tribes. These reports continued into the 20th century, with several Western researchers and explorers claiming to have seen the creature firsthand. Sightings are most common in the Kerinci Seblat National Park, a remote and mountainous area covered in dense rainforest. While no physical evidence of the Orang Pendek has been found, researchers have discovered footprints and other signs of the creature's presence, leading some to speculate that it could be a surviving population of an ancient hominid species, such as Homo floresiensis, the so-called "Hobbit" species discovered on the nearby island of Flores.

In Australia, the Yowie is the continent's answer to Bigfoot. Descriptions of the Yowie vary, but it is generally depicted as a large, hairy, bipedal creature, standing between six and ten feet tall, with ape-like features and long arms. The Yowie is said to inhabit the remote wilderness areas of Australia, particularly in the eastern states of New South Wales and Queensland. Sightings of the Yowie have been reported by both Indigenous Australians and European settlers, with the creature often described as shy and elusive, but also capable of aggression if provoked.

Like Bigfoot, the Yowie has become a part of Australian pop culture, with sightings and encounters regularly reported in the media. However, despite the popularity of the legend, physical evidence of the Yowie remains elusive. Researchers and cryptozoologists continue to investigate reports of the creature, with some drawing parallels between the Yowie and Bigfoot, suggesting that both could be members of an unknown species of hominid that once roamed the Earth.

In Russia and parts of Central Asia, stories of the Almasty, also known as the Russian Bigfoot or wild man, have been passed down through generations. The Almasty is typically described as a large, hairy, human-like creature, similar to Bigfoot, but with more pronounced facial features and sometimes a more human-like appearance. Reports of the Almasty are concentrated in the remote and mountainous regions of the Caucasus and Pamir mountains, as well as in Siberia.

Russian cryptozoologists have taken a particular interest in the Almasty, with several expeditions launched to search for the creature in the 20th and 21st centuries. Some researchers have speculated that the Almasty could be a surviving population of Neanderthals or Denisovans, ancient human relatives that lived alongside modern humans tens of thousands of years ago. While no conclusive evidence has been found, the search for the Almasty continues, with occasional sightings and footprints reported by locals and researchers alike.

In Africa, the legend of the Kikomba or Kikundak is found in the dense rainforests of the Congo Basin. The Kikomba is said to be a large, bipedal creature, similar to Bigfoot, but with more ape-like features. Local tribes believe the Kikomba to be a guardian of the forest, with the ability to vanish into the trees when threatened. Sightings of the Kikomba are rare, but the creature remains an important figure in the folklore of the region.

One of the most intriguing aspects of these global wild man legends is the consistency in their descriptions. Whether it's the Yeti in the Himalayas, the Orang Pendek in Sumatra, or the Yowie in Australia, these creatures are often described as large, bipedal, and covered in hair. They are elusive, intelligent, and capable of avoiding human detection, living in remote, often inhospitable areas. This consistency has led some researchers to speculate that the legends of these creatures may be rooted in a shared human experience, perhaps a distant memory of ancient hominids that once roamed the Earth alongside modern humans.

While no definitive evidence of these creatures has been found, their presence in global folklore suggests that the idea of the wild man is deeply ingrained in the human psyche. Whether as guardians of the wilderness, symbols of the unknown, or reminders of our own primal origins, these creatures continue to captivate the imagination of people around the world. And as the search for Bigfoot and its global counterparts continues, the wild men of the world remain powerful symbols of the untamed and the unexplained.

Chapter 16

As the legend of Bigfoot continues to captivate the imagination of millions, the future of Bigfoot research is evolving. Advances in technology, a growing interest in cryptozoology, and the global spread of information have created new opportunities for researchers and enthusiasts alike. While the mystery of Bigfoot remains unsolved, the quest to uncover the truth is far from over. In this chapter, we'll explore the future of Bigfoot research, the new tools and approaches being developed, and what lies ahead for those seeking to solve one of the greatest cryptozoological puzzles of our time.

One of the most promising areas of Bigfoot research lies in the development of new technologies that can aid in the search for elusive creatures. As discussed in earlier chapters, trail cameras, thermal imaging, and eDNA) sampling have already been employed by researchers with some success, but new tools and innovations are on the horizon that could further revolutionize the field.

Drones, for example, are becoming an increasingly important tool in wildlife research, allowing scientists and cryptozoologists to cover large areas of rugged, inaccessible terrain quickly and efficiently. Equipped with high-definition cameras and thermal imaging technology, drones can capture images and heat signatures from the air, offering a new perspective on potential Bigfoot habitats. As drone technology continues to improve, it's likely that drones will play an even greater role in future Bigfoot expeditions, providing researchers with a bird's-eye view of the remote forests and mountains where Bigfoot is said to reside.

In addition to drones, artificial intelligence (AI) is also being explored as a tool for analyzing data related to Bigfoot sightings. With thousands of sightings reported each year, it can be difficult for researchers to sift through the data and identify patterns or anomalies. AI programs can help by analyzing large datasets, identifying trends in sightings, and flagging areas where Bigfoot activity might be more likely. This technology could allow researchers to focus their efforts on specific regions, improving the chances of capturing definitive evidence.

The use of AI extends beyond data analysis. Some researchers are developing machine learning algorithms that can analyze video and audio recordings for potential Bigfoot evidence. For example, AI could be used to scan hours of trail camera footage or audio recordings, searching for movement patterns, heat signatures, or vocalizations that might be indicative of Bigfoot. This could save researchers valuable time and increase the chances of detecting the creature in its natural environment.

Another promising development in Bigfoot research is the continued advancement of DNA analysis techniques. While eDNA has shown potential in detecting rare species through environmental samples, newer methods of DNA extraction and sequencing are becoming available that could make it easier to identify unknown species. Researchers are now able to extract DNA from even the smallest traces of biological material, such as hair, skin cells, or feces, and compare the genetic material to known species in global DNA databases. If a sample were to return an unknown or anomalous result, it could point to the existence of an undiscovered species, such as Bigfoot.

Additionally, the growing interest in citizen science is likely to shape the future of Bigfoot research.

With the rise of smartphones, apps, and social media, more people than ever before have the tools to document their encounters with potential cryptids. Apps like iNaturalist, which allow users to record wildlife sightings and share them with the scientific community, could be adapted to include reports of cryptids like Bigfoot. This would create a centralized database of sightings that researchers could analyze in real time, helping to identify potential hotspots for Bigfoot activity.

Citizen science has already played a significant role in the study of other cryptids and rare species. For example, the discovery of the Saola in Vietnam was made possible in part by local villagers reporting sightings to scientists.

Similarly, amateur birdwatchers have contributed to the discovery of new bird species in remote areas. As more people engage with cryptozoology through citizen science, the chances of finding new evidence of Bigfoot could increase.

Of course, the future of Bigfoot research will also depend on continued funding and institutional support. While cryptozoology has traditionally occupied a fringe position in the scientific community, there are signs that this is beginning to change. Some universities and research institutions have started to take cryptozoology more seriously, recognizing the value of studying unknown species and the importance of preserving wilderness areas where such creatures might reside. As more scientists become open to the possibility of undiscovered species, cryptozoology could gain greater legitimacy and attract more funding for research.

The media will also continue to play a role in shaping the future of Bigfoot research. Television shows like "Finding Bigfoot," documentaries, and podcasts dedicated to cryptids have kept the legend of Bigfoot alive in the public imagination. These platforms not only entertain but also educate viewers about the scientific methods used to investigate cryptids, sparking interest in the field and inspiring a new generation of researchers. As the popularity of cryptozoology grows, it's likely that we will see more documentaries, films, and books dedicated to Bigfoot, further fueling the public's fascination with the creature.

Despite the technological advancements and growing interest in cryptozoology, the search for Bigfoot still faces significant challenges. The creature, if it exists, is elusive, living in some of the most remote and difficult-to-access areas of the world. Even with the best tools and technology available, capturing definitive proof of Bigfoot's existence will be no easy task. The search will require patience, persistence, and a willingness to accept that the mystery may never be fully solved.

However, for many researchers and enthusiasts, the search itself is just as important as the discovery. The quest for Bigfoot represents a deeper human desire to explore the unknown, to challenge the limits of our knowledge, and to preserve the wild places that still exist in the world. As long as there are unexplored forests and unanswered questions, the search for Bigfoot will continue, driven by the hope that one day, the truth will be revealed.

Whether or not Bigfoot is ever found, the pursuit of the creature has already contributed to our understanding of wildlife, wilderness conservation, and the boundaries of scientific knowledge. The technologies and methods developed in the search for Bigfoot have applications beyond cryptozoology, helping researchers study endangered species, monitor ecosystems, and track elusive animals. In this sense, Bigfoot research is about more than just finding a creature—it's about pushing the boundaries of exploration and discovery, both in the natural world and within ourselves.

As we look to the future, one thing is certain: the legend of Bigfoot will continue to inspire curiosity, debate, and adventure for generations to come. Whether as a symbol of the unknown, a guardian of the wilderness, or a living creature waiting to be discovered, Bigfoot remains one of the most enduring mysteries of our time. And as long as there are those willing to search, the possibility of finding Bigfoot will remain, tantalizingly, just beyond the horizon.

Chapter 17

The legend of Bigfoot has endured for centuries, captivating the imaginations of people across cultures, generations, and belief systems. Despite the lack of definitive evidence, the creature continues to inspire countless books, television shows, documentaries, and even scientific research. But what is it about Bigfoot that keeps the legend alive? In this chapter, we'll explore the psychology behind the belief in Bigfoot, examining the cognitive, emotional, and cultural factors that drive people to search for—and sometimes believe in—the existence of this elusive creature.

At the heart of the Bigfoot phenomenon is a fundamental aspect of human psychology: our tendency to seek meaning and patterns in the world around us. Humans are naturally wired to detect patterns and make connections between seemingly unrelated events. This ability has been crucial for survival throughout our evolutionary history, helping us to identify predators, find food, and navigate complex social environments. However, this same cognitive tendency can sometimes lead us to perceive patterns where none exist, such as seeing a blurry figure in the woods and interpreting it as a Bigfoot sighting.

This phenomenon, known as pareidolia, plays a significant role in many cryptid sightings. Pareidolia is the brain's tendency to impose familiar shapes or images onto random stimuli, such as seeing faces in clouds or interpreting a shadowy figure as Bigfoot. In the wilderness, where unfamiliar sights and sounds abound, it's easy to see how a brief glimpse of a large animal or a rustling in the bushes could be misinterpreted as evidence of something more mysterious. Our brains are quick to fill in the gaps, especially when we're in unfamiliar or uncertain environments.

Another important psychological factor that contributes to the belief in Bigfoot is cognitive bias. One of the most well-known forms of cognitive bias is confirmation bias, the tendency to seek out and interpret information in ways that confirm our existing beliefs. For those who already believe in the existence of Bigfoot, it's natural to interpret ambiguous evidence, such as footprints or vocalizations, as proof of the creature's presence. Similarly, those who are skeptical may be more likely to dismiss the same evidence as hoaxes or misidentifications. In this way, cognitive biases can reinforce existing beliefs and make it more difficult for individuals to objectively evaluate new information.

In addition to cognitive biases, emotional factors play a powerful role in shaping belief in Bigfoot. The search for Bigfoot is often driven by a sense of wonder and curiosity—a desire to explore the unknown and discover something extraordinary. For many people, the idea that there are still mysteries left to uncover in the natural world is deeply appealing, especially in an age where much of the world has been mapped, studied, and explained by science. Bigfoot represents the possibility that there are still wild, untamed places left to explore, and that the world is not as fully understood as we might believe.

This emotional connection to Bigfoot is often rooted in a broader sense of nostalgia for a time when the world felt larger and more mysterious. For some, the search for Bigfoot is a way of reconnecting with nature and rediscovering a sense of adventure that has been lost in the modern, technology-driven world. The creature embodies the idea of the wilderness as a place of mystery and danger, where anything is possible. In this sense, Bigfoot serves as a symbol of the unknown—a reminder that there are still frontiers left to explore, both in the physical world and in the human imagination.

Cultural factors also play a significant role in the enduring belief in Bigfoot. The legend of Bigfoot is deeply embedded in North American folklore, with Indigenous tribes having told stories of large, hairy creatures for centuries. These stories, passed down through generations, have shaped the modern perception of Bigfoot as a guardian of the forest and a creature that exists beyond the reach of human civilization. Similar stories exist in cultures around the world, from the Yeti in the Himalayas to the Orang Pendek in Indonesia, suggesting that the idea of a mysterious wild man is a common theme in human mythology.

In modern times, the media has played a key role in perpetuating the legend of Bigfoot. Television shows like "Finding Bigfoot" and documentaries about cryptids have kept the creature in the public eye, introducing new generations to the legend and sparking renewed interest in the search. The portrayal of Bigfoot in popular culture has also evolved over time, with the creature being depicted in a variety of ways, from a terrifying, dangerous predator to a misunderstood, gentle giant. This flexibility in how Bigfoot is portrayed allows the legend to resonate with a wide range of audiences, each interpreting the creature in their own way.

The role of the internet in shaping belief in Bigfoot cannot be overstated. Online forums, social media platforms, and websites dedicated to cryptozoology have created a global community of Bigfoot enthusiasts, allowing people to share sightings, evidence, and theories in real time. The internet has democratized the search for Bigfoot, making it easier for amateur researchers and enthusiasts to participate in the quest for discovery. However, it has also contributed to the spread of hoaxes and misinformation, as digital editing tools and the viral nature of social media make it easier to fabricate evidence and deceive others.

Despite the challenges posed by hoaxes and skepticism, the belief in Bigfoot endures in part because it fulfills a deep psychological need for mystery and the unknown. Humans are naturally curious creatures, and we are drawn to the idea that there are still secrets left to uncover in the world. Bigfoot represents the possibility that not everything has been explained by science, and that there are still wonders waiting to be discovered. For many, the search for Bigfoot is not just about finding a creature—it's about finding meaning in a world that can often feel predictable and fully mapped.

Another psychological factor that contributes to the belief in Bigfoot is the human tendency toward magical thinking. Magical thinking is the belief that certain events or phenomena are caused by supernatural forces or that the world is governed by unseen powers. In the case of Bigfoot, some people may believe that the creature possesses supernatural abilities, such as the power to vanish into thin air or evade detection by cameras and other technology. This belief in Bigfoot as a magical or spiritual being adds another layer of mystery to the legend, making it even more difficult for skeptics to disprove.

Ultimately, the belief in Bigfoot is not just about evidence or logic—it's about hope. Hope that there are still mysteries left to explore, hope that the natural world is wilder and more untamed than we realize, and hope that we might one day discover something extraordinary. For those who believe, the search for Bigfoot is a way of holding on to that sense of wonder and possibility, even in the face of skepticism and doubt.

As long as people continue to seek meaning in the unknown, the legend of Bigfoot will endure. Whether or not the creature is ever found, its place in the human imagination is secure, serving as a symbol of the wild, the mysterious, and the untamed. And as long as there are those willing to venture into the wilderness in search of the creature, the legend of Bigfoot will live on, captivating the hearts and minds of those who dare to believe in the possibility of the unknown.

Chapter 18

The search for Bigfoot, like the pursuit of other cryptids, is more than a simple quest for evidence. It taps into a deeper human fascination with the unknown and speaks to our enduring desire to explore the boundaries of what we understand. Cryptozoology, the study of hidden or unknown animals, plays a unique role in society and psychology by feeding our collective imagination and offering a sense of wonder in an increasingly mapped and explained world. In this chapter, we'll explore the social and psychological forces that drive people to seek the unknown, and how the study of cryptids like Bigfoot connects us to something larger than ourselves.

One of the core reasons for humanity's interest in cryptids is rooted in our evolutionary history. As early humans, we needed to be constantly aware of our environment, identifying potential dangers and seeking out resources in unfamiliar landscapes. This evolutionary impulse to explore and understand the unknown persists in modern times, though we are no longer hunting for food or evading predators. Instead, we satisfy this impulse through exploration of new frontiers, whether it be outer space, the depths of the ocean, or the uncharted wilderness where creatures like Bigfoot are said to roam.

The drive to seek out the unknown is closely tied to a fundamental human need for curiosity and discovery. Curiosity has been the engine behind some of humanity's greatest achievements, from scientific breakthroughs to geographic explorations. It is also the psychological basis for why we are drawn to mysteries like Bigfoot. The idea that there could be undiscovered creatures living among us fuels our sense of wonder and pushes us to explore the world more deeply.

But it's not just curiosity that keeps us searching for Bigfoot and other cryptids—it's also a desire for meaning. In a world that can often feel predictable and overly rational, the idea of something unknown and mysterious offers a break from the mundane. Bigfoot represents the possibility that there is more to the world than what science and technology can explain. For many, this belief provides a sense of hope that the world is still full of surprises and that nature holds more secrets than we have yet discovered.

Cryptozoology also taps into a collective longing for a connection to nature. In modern society, many people feel disconnected from the natural world, spending most of their time in urban environments and rarely venturing into wilderness areas. The search for cryptids like Bigfoot offers a way to reconnect with nature and rediscover the awe-inspiring power of the wilderness. Bigfoot is often seen as a symbol of the wild, an embodiment of the untamed, mysterious forces of nature that exist beyond human control.

This sense of connection to nature is further reinforced by the cultural role of Bigfoot and other cryptids. Stories of wild men or hidden creatures have been passed down for generations in many cultures, from the Indigenous peoples of North America to the Himalayan Sherpas who speak of the Yeti. These stories often serve as a reminder of the power and mystery of the natural world, cautioning against human hubris and encouraging respect for the wilderness. In this way, the belief in cryptids functions as a form of ecological storytelling, reminding us of the importance of preserving the wild places where these creatures might still exist.

From a psychological perspective, belief in cryptids can also be seen as a form of magical thinking.

Magical thinking refers to the human tendency to believe in forces or beings that exist outside the realm of science or logic. This type of thinking is deeply ingrained in human psychology, dating back to ancient times when people believed in gods, spirits, and other supernatural entities. While modern science has explained many of the mysteries that once puzzled humans, the belief in cryptids persists, offering a modern form of magical thinking that allows people to believe in the possibility of the unknown.

.

Magical thinking is not inherently irrational—it often serves as a way for people to cope with uncertainty and make sense of experiences that fall outside of their usual understanding of the world. For those who believe in Bigfoot, the creature represents the possibility that not everything can be explained by science, and that there are still forces at work in the world that we have yet to comprehend. This belief can be comforting, especially in a world that often feels increasingly controlled by technology and empirical data.

In addition to its psychological appeal, cryptozoology also fulfills important social functions. The search for cryptids like Bigfoot brings people together, fostering a sense of community and shared purpose. Cryptozoology conventions, online forums, and field expeditions offer spaces where people can connect with others who share their passion for the unknown. These communities provide validation and support for believers, helping them feel like they are part of something larger than themselves. In this way, cryptozoology functions as a form of social bonding, creating a collective identity around the pursuit of discovery.

The social role of cryptozoology is further reinforced by the media's portrayal of cryptids like

Bigfoot. Television shows, documentaries, and podcasts dedicated to the search for cryptids have brought the phenomenon into the mainstream, attracting a global audience. These media platforms not only entertain but also encourage people to participate in the search, whether by reporting sightings, joining expeditions, or simply sharing theories online.

The rise of social media has also played a significant role in spreading cryptid stories, allowing enthusiasts to share evidence, debate theories, and build communities in real time.

While belief in cryptids may seem fringe to some, it speaks to a broader human desire for meaning and connection. Cryptozoology offers a way for people to explore the mysteries of the world while also satisfying psychological and social needs. For many, the search for Bigfoot is about more than finding a creature—it's about finding wonder, hope, and a sense of belonging in a world that often feels predictable and overly explained.

As long as people continue to seek the unknown, cryptozoology will remain a vital part of human culture. Whether or not Bigfoot is ever found, the search for the creature will continue to inspire curiosity, imagination, and exploration. In this way, cryptozoology serves as a reminder that the world is still full of mysteries waiting to be uncovered, and that the human spirit of discovery is as strong as ever.

Chapter 19

While Bigfoot is perhaps the most famous cryptid, the field of cryptozoology extends far beyond this single creature. Cryptozoologists around the world are dedicated to studying and searching for a wide variety of hidden or unknown animals, ranging from lake monsters to mysterious sea creatures to humanoid beings. As technology advances and our understanding of the natural world deepens, the future of cryptozoology holds exciting possibilities for discovery. In this chapter, we will explore how cryptozoology is evolving, the role of technology in future research, and the potential for uncovering new species that may reshape our understanding of the animal kingdom.

Cryptozoology has long been viewed as a fringe science, often dismissed by mainstream scientists as pseudoscience due to the lack of conclusive evidence for many of the creatures it studies. However, recent discoveries of previously unknown species and the development of new research tools have begun to shift the perception of cryptozoology. While the field still faces significant skepticism, it is becoming increasingly clear that there are still many mysteries to be uncovered in the natural world, and cryptozoologists may be playing a key role in pushing the boundaries of scientific exploration.

One of the most significant ways cryptozoology is evolving is through the use of advanced technology. Tools such as eDNA, drones, and thermal imaging have already been employed in the search for cryptids like Bigfoot, but these technologies are being refined and expanded for use in a wider range of environments and creatures. For example, eDNA sampling has been used in aquatic environments to detect rare or elusive species, such as giant squids or endangered fish species. This same technology could be used to search for lake monsters or other unknown aquatic creatures that have long been part of folklore.

Drones are another tool that holds promise for future cryptozoological research. The ability to cover large, remote areas with high-definition cameras and thermal sensors has made it easier for researchers to monitor difficult-to-access habitats, such as dense forests, mountain ranges, and oceanic regions. In the search for creatures like Bigfoot, the Loch Ness Monster, or even more elusive cryptids like the Mokele-Mbembe—a dinosaur-like creature said to inhabit the rivers of the Congo Basin—drones offer the potential to explore areas that would otherwise be impossible for human researchers to reach.

AI is also emerging as a valuable tool in cryptozoology. AI algorithms can be used to analyze massive datasets, such as footage from trail cameras, satellite images, or acoustic recordings, to identify patterns or anomalies that could indicate the presence of unknown species. This technology has already been applied in wildlife conservation efforts, such as tracking endangered species or identifying illegal poaching activities. In the context of cryptozoology, AI could help researchers sift through large volumes of data more efficiently, improving the chances of finding compelling evidence.

Another important development in cryptozoology is the growing recognition of local knowledge and indigenous stories as valuable sources of information. For centuries, indigenous peoples around the world have told stories of mysterious creatures, many of which bear striking similarities to the cryptids studied by modern cryptozoologists. In some cases, these creatures have been dismissed by Western scientists as myths or legends, but increasingly, cryptozoologists are recognizing the importance of listening to and collaborating with local communities in their search for new species.

For example, the discovery of the Saola in Vietnam—a rare species of antelope that was unknown to science until 1992—was made possible in part by the knowledge and reports of local villagers. Similarly, the Orang Pendek, a small, bipedal cryptid said to inhabit the forests of Sumatra, has been the subject of study by cryptozoologists who have worked closely with indigenous tribes to gather evidence and conduct field research. By respecting and incorporating indigenous knowledge, cryptozoologists can gain valuable insights into the habits, behavior, and habitats of potential cryptids, increasing the likelihood of making new discoveries.

As cryptozoology moves forward, the integration of traditional scientific methods with emerging technologies and local knowledge will be crucial to its success. The discovery of new species is not only a goal for cryptozoologists but also a critical aspect of wildlife conservation. Many of the environments where cryptids are said to live are some of the most biodiverse and fragile ecosystems on Earth. The search for unknown species can serve as a powerful motivator for preserving these habitats, as well as a way to raise awareness about the importance of conservation efforts.

However, the future of cryptozoology will also require a greater degree of scientific rigor. One of the challenges the field has faced is its association with hoaxes, sensationalism, and a lack of peer-reviewed research. For cryptozoology to gain greater credibility, researchers will need to adopt stricter standards for evidence collection, data analysis, and publication. This includes conducting controlled experiments, submitting findings to peer-reviewed journals, and maintaining transparency in the research process. By adhering to these standards, cryptozoologists can help bridge the gap between their field and the broader scientific community.

Cryptozoology also holds potential for exciting new discoveries beyond the search for specific cryptids. In recent years, the discovery of new species in remote and previously unexplored areas has demonstrated that the natural world is far from fully understood. In the last few decades alone, scientists have discovered species such as the Tapanuli orangutan in Sumatra, the psychedelic frogfish in Indonesia, and Omura's whale off the coast of Madagascar. These discoveries remind us that the world is still full of surprises and that new species can be found in the most unexpected places.

While many of the creatures studied by cryptozoologists may never be found, the search itself contributes to our understanding of biodiversity and the complexities of ecosystems. The study of cryptids encourages a broader exploration of the natural world and challenges us to think more creatively about the limits of scientific knowledge. It also encourages a sense of wonder and curiosity, reminding us that there is still much left to discover about the planet we call home.

As we look to the future of cryptozoology, it's clear that the field is evolving in exciting ways. Advances in technology, growing collaboration with indigenous communities, and a commitment to scientific rigor are all contributing to a new era of exploration and discovery. Whether or not creatures like Bigfoot, the Loch Ness Monster, or the Yeti are ever found, the search for these cryptids will continue to inspire a deeper understanding of the natural world and the many mysteries it holds.

Ultimately, cryptozoology is about more than just finding hidden creatures—it's about pushing the boundaries of what we know and remaining open to the possibility that there is still much left to learn. As new technologies emerge and researchers continue to explore the farthest reaches of the Earth, the future of cryptozoology holds immense potential for discovery, both in the natural world and in our understanding of the unknown.

Chapter 20

During the Vietnam War, tales of encounters with mysterious, ape-like creatures—commonly referred to as "rock apes"—became a notable part of soldier lore. My brother-in-law, a veteran of the conflict, vividly recalled his experiences with these enigmatic beings. His accounts, primarily centered on the areas around Firebase St. Barbara and Nui Ba Den, offer a unique glimpse into a world where the boundaries between the known and the unknown blurred amid the chaos of war. These creatures, both awe-inspiring and unsettling, were more than mere stories. They seemed tangible, embodied in the jungles' shadows and echoes, and their presence left a lasting impression on those who encountered them.

The rock apes, as he described them, were unlike any known primates in the region. Standing upright when provoked or agitated, they ranged from approximately 5′11″ to over 6 feet tall, with broad shoulders and heavily muscled frames. Their fur, varying in color from deep black to reddish-brown, allowed them to blend seamlessly with their surroundings, adding to their elusive nature. These creatures appeared far more robust than macaques or gibbons, with an estimated weight well exceeding 200 pounds. My brother-in-law was convinced they were a distinct species—intelligent, strong, and capable of behaviors that bordered on human-like.

One of the most remarkable traits of the rock apes was their complex and seemingly deliberate behavior. They would often rise to their full height when interacting with or confronting humans, a display that seemed to assert dominance or evaluate potential threats. Unlike typical wildlife, which tended to avoid humans, the rock apes exhibited a mix of curiosity and territorial aggression. They communicated through a range of vocalizations, from loud grunts to piercing screams that carried across the jungle, serving both as warnings and declarations of their presence. These sounds often punctuated the stillness of the night, startling soldiers and leaving them with an acute sense of vulnerability.

The rock apes' most notable behavior, and the one that earned them their name, was their propensity to hurl stones. My brother-in-law marveled at their precision and power. Rocks ranging from small pebbles to softball-sized projectiles would fly with alarming accuracy, often pelting soldiers stationed near the edge of the jungle. He estimated that the apes could launch these rocks over 50 yards with impressive force. This behavior, while seemingly aggressive, also seemed calculated—less an attempt to inflict harm and more a demonstration of their capabilities and dominance.

The creatures' nightly raids on the base's dump area revealed another facet of their intelligence and adaptability. Drawn by the promise of easy pickings, the rock apes would sift through discarded food scraps and other refuse, carefully selecting items of interest. This scavenging behavior became so routine that soldiers began to anticipate their visits, often observing them from a safe distance. Some, however, were less cautious.

One such individual was a soldier nicknamed "Chief," an American Indian friend of my brother-in-law, who took his interactions with the rock apes to an unusual and daring level. Chief, armed with rocks of his own, would venture near the dump under cover of darkness and provoke the creatures into their characteristic rock-throwing game. The exchanges were intense, with rocks flying back and forth until Chief retreated, battered but laughing. To him, it was a strange, nightly ritual—a connection of sorts between humans and these wild beings. My brother-in-law thought him reckless, but the tales of Chief's antics became part of the growing lore.

Perhaps the most unnerving encounter came when my brother-in-law witnessed a large group of rock apes, numbering between ten and fifteen individuals, emerging from the jungle near Nui Ba Den. The sheer number of creatures was a stark contrast to the smaller groups or solitary individuals he had previously encountered. Among them were smaller figures, possibly juveniles or females, though no very young apes were observed. This sighting reinforced his belief that the rock apes were social creatures, moving in loosely organized packs. The destruction and noise of the war, he speculated, may have disrupted their habitats and driven them closer to human activity, forcing these rare encounters.

The areas around Nui Ba Den and Firebase St. Barbara weren't the only locations where soldiers reported such sightings. Accounts of rock apes extended to the Mekong Delta, the Cu Chi Tunnels, and Tay Ninh Province, though with some variations. In the swampy expanses of the Delta, reports of small groups of upright, ape-like creatures were rarer, as the terrain seemed less suited to their size and mobility. Near the Cu Chi Tunnels, soldiers described being pelted with stones in dense forested areas, but sightings were fleeting, with shadows vanishing before they could be identified.

The terrain around Tay Ninh Province, with its dense vegetation and proximity to the Cambodian border, seemed more conducive to the rock apes' movements. Soldiers in this region often described catching glimpses of large, bipedal figures moving through the underbrush, always just out of reach. While skeptics argued that many of these encounters could be attributed to misidentifications of known species, such as macaques, the physical descriptions provided by witnesses suggested something far more extraordinary.

Skeptics also pointed to the psychological stress of war as a factor in these stories. The relentless tension, isolation, and constant threat of danger could heighten perceptions and lead to exaggerations. Yet, my brother-in-law maintained that what he and his comrades witnessed was real, distinct from any wildlife they had encountered before. He was familiar with the region's animals, and the rock apes' size, behavior, and intelligence set them apart.

Reflecting on these encounters, it's easy to see how the rock apes became a symbol of the untamed wilderness of Vietnam. They were not merely animals but embodiments of the jungle's mysteries—a reminder of the unknown forces that persisted even in the midst of modern warfare. Whether they were an undiscovered species, displaced by the chaos of war, or a manifestation of deeper cultural and psychological currents, their stories endure as part of the collective memory of those who served.

The rock apes leave us with questions that extend beyond the bounds of science or myth. Were they a relic of some lost lineage, hidden in the uncharted reaches of Vietnam's jungles? Or were they a symbol of humanity's eternal confrontation with the wild, the enigmatic, and the untamed? Whatever the answer, these creatures live on in the stories of the soldiers who encountered them, a testament to the enduring mystery of the natural world.

Chapter 21

Bigfoot researchers and investigators, both amateur and professional, dedicate countless hours to scouring forests, mountains, and swamps for signs of the elusive creature. Yet, despite the passion and dedication, the field of Bigfoot research is rife with challenges, many of which stem from human error. From well-meaning weekend investigators to seasoned experts, mistakes can hinder the credibility of findings and leave Bigfoot's true existence clouded in doubt.

In this chapter, we will explore some of the most common mistakes made by Bigfoot investigators, the unintended consequences of well-intentioned research, and ways to eliminate or mitigate these issues.

1. Tainted Evidence: Tree Knocks, Howls, and Cross-Contamination

One of the most popular methods used by Bigfoot investigators involves mimicking known behaviors attributed to the creature, most notably tree knocks and howls. These sounds, supposedly used by Bigfoot to communicate over long distances, are often recreated by investigators in the hopes of eliciting a response. While this approach has gained popularity, it has unintentionally led to a serious problem: the contamination of potential evidence.

Imagine this scenario: An investigator hears a distant tree knock, responds with a knock of their own, and soon hears another. They return home, convinced they've had an interaction with a Sasquatch. However, there's a glaring flaw. How many other Bigfoot researchers were in the same area, possibly miles away, performing similar techniques? It's impossible to determine whether the sounds came from a genuine Bigfoot or just another well-meaning investigator.

This issue is exacerbated during large, organized expeditions where multiple groups operate in the same general area. Even more problematic are vocalizations, as enthusiasts love to mimic reported Bigfoot howls. While the goal is to spark a response, this practice often results in false positives. The howls heard deep in the woods might simply be echoes of another research team on the other side of the ridge. These uncontrolled and unsystematic methods create a chaotic field where evidence, though exciting, often traces back to humans rather than Sasquatch.

2. The Challenge of DNA Collection and Cross-Contamination

The advent of DNA testing was supposed to be a game-changer in the hunt for Bigfoot. Hair samples, eDNA, and scat analysis have become the gold standard for attempting to scientifically prove the creature's existence. However, this has brought about its own set of challenges, particularly in the collection and preservation of samples.

One of the biggest issues is cross-contamination. Human DNA can easily contaminate samples through touch, improper storage, or accidental exposure to other biological materials. Investigators often collect hair samples with bare hands, use makeshift containers, or expose samples to the environment before sending them to laboratories. A single stray hair from the collector, a microscopic fiber from their clothing, or even a few skin cells can render an otherwise promising sample scientifically useless.

Another mistake involves the misidentification of samples. Hair or scat found in the wilderness is often presumed to be from Bigfoot, but upon analysis, many such samples turn out to be from known animals, such as bear, deer, or even humans. In the excitement of discovery, many investigators skip the thorough vetting process, which could prevent these false identifications from being publicized as "breakthrough" evidence.

3. Ill-Conceived Tactics to Attract Bigfoot

To lure Bigfoot into the open, some investigators employ unconventional or even absurd tactics, often without fully considering their effectiveness or consequences. There have been reports of people leaving out food, playing music, and even dressing in costumes to interact with the creature.

Some tactics, like food baiting, attract known wildlife such as bear, raccoons, or deer, leading to dangerous or misleading situations. Worse, baiting can alter the behavior of local wildlife populations, conditioning them to associate human presence with easy food sources. This can skew the ecological balance of an area and even put other humans at risk.

Other ideas are far more eccentric. Attempts to attract Bigfoot using pheromone chips, primal recordings of ape calls, or gifting shiny trinkets like mirrors and toys often border on pseudoscience. While these efforts are rooted in good intentions, they lack any real basis in ethological research (the study of animal behavior) and risk undermining the credibility of Bigfoot research as a whole.

4. The Influence of Confirmation Bias

One of the most insidious challenges faced by Bigfoot researchers is confirmation bias. Many enter the field already convinced Bigfoot exists, and as a result, they interpret every ambiguous sound, movement, or shape as evidence of the creature. A twig snapping in the night must be Bigfoot; rustling leaves? Bigfoot on the move.

While optimism and belief are powerful motivators, they can lead to faulty conclusions. Confirmation bias clouds judgment, making investigators more likely to overlook alternative explanations, such as wind, other animals, or even humans in the woods.

Scientific research relies on objectivity, but when researchers are too eager for validation, their evidence—however promising—loses credibility.

5. Lack of Scientific Procedure in Field Research

Another major issue plaguing Bigfoot investigations is the lack of consistent scientific procedure. Many well-meaning enthusiasts are passionate but fail to employ proper methodologies during research. Data collection is often unsystematic, with few controls to rule out other explanations for observed phenomena.

For instance, investigators frequently rely on anecdotal evidence, such as eyewitness testimony or personal experiences, without corroborating these claims with hard data. While personal encounters are compelling, they cannot stand alone as scientific proof.

Many research teams also neglect long-term, controlled studies in favor of short expeditions that produce tantalizing but inconclusive results. Furthermore, investigators often lack training in wildlife biology, ecology, or forensic science—skills essential for interpreting tracks, hair, scat, or sounds.

Despite challenges, concrete steps can be taken to improve Bigfoot research:

1. Coordination Between Researchers

Establish communication channels between research teams to prevent cross-contamination of tree knocks and howls.

Use a shared database to log activities, ensuring responses aren't inadvertently caused by nearby investigators.

2. Proper DNA Collection Protocols

Use gloves, sterile containers, and proper chain-of-custody procedures when collecting samples.

Partner with professional forensic scientists or biologists for analysis.

3. Ethologically Sound Attraction Techniques

Base methods on scientifically sound ideas.

Minimize baiting to avoid disrupting local ecosystems.

4. Mitigating Confirmation Bias

Document alternative explanations for findings.

Corroborate evidence with multiple data points, such as tracks, hair, and sounds.

5. Scientific Rigor

Adopt standardized methods for data collection.

Work with professionals from related fields, such as anthropology and forensic science.

Addressing these shortcomings can strengthen Bigfoot investigators' credibility and improve their chances of uncovering definitive evidence. In the next chapter, we'll explore how professional-grade tools and dedicated funding could revolutionize Bigfoot research.

Chapter 22

Despite the best intentions, Bigfoot research faces numerous challenges that often undermine its credibility. From tainted evidence caused by tree knocks and howls made by other researchers to cross-contamination of DNA samples and ill-conceived methods for attracting the creature, the road to proving Bigfoot's existence is both complicated and fraught with setbacks. To elevate the field, Bigfoot investigators must embrace rigorous scientific methods, establish better coordination among teams, and apply a critical, objective lens when analyzing potential evidence. Only by addressing these shortcomings can we hope to separate human error from genuine evidence and, perhaps one day, uncover the truth about whether Bigfoot is indeed hiding in plain sight.

I enjoy watching "Expedition Bigfoot", a program airing on the Discovery Channel. The members of the team place themselves in the field for several days and weeks, searching for evidence of an undiscovered North American great ape or a possible relic hominin.

While there have been some misses when it comes to shows dedicated to Bigfoot research, that's to be expected. What matters most is a focus on Bigfoot as a flesh-and-blood creature.

Unfortunately, many programs veer into theories involving the paranormal, extraterrestrials, or extra-dimensional explanations to satisfy fringe elements of their audience. Personal perception becomes truth: if someone is expecting answers outside of the physical realm, they're likely to interpret encounters through that lens, slotting Bigfoot into their belief system. However, the simplest explanation is often the most likely one.

North American Indigenous people who believe Bigfoot to be a spiritual being carry an honored tradition passed down through generations. Their stories and beliefs deserve respect, especially as they often intertwine the spiritual with the physical. Many Indigenous accounts describe Bigfoot as both a mystical figure and a real, physical entity coexisting in the wilderness.

One can only hope that a billionaire might someday step up to the plate and fund a state-of-the-art expedition for Bigfoot research. Imagine a six-month field study equipped with the latest technology and a team of qualified experts. This dream team would include specialists in physical anthropology, primatology, cultural anthropology, veterinary science, and support staff for audio, video, linguistic, and historical research.

Let's delve deeper into the requirements for such an expedition and outline the equipment and expertise that would be needed for success:

Proper Bigfoot research demands specialized tools to enhance observation, documentation, and sample preservation in remote environments. Here's a list of recommended equipment for a state-of-the-art expedition:

High-resolution night vision cameras and devices like the GoPro HERO9 Black offer 5K video capabilities and rugged durability, enabling the capture of clear footage.

Thermal Imaging Cameras–Crucial for detecting heat signatures in low-light environments.

Professional DSLR or Mirrorless Cameras Equipped with zoom lenses to capture detailed images from a distance.

High-Quality Audio Recorders–To document potential vocalizations or other unique sounds.

SIONYX Aurora Series Full-Color Night Vision Cameras–Provide full-color visibility in low-light conditions for enhanced nocturnal observation.

Rugged Durable, high-speed laptops designed to withstand harsh field conditions for data analysis and storage.

Satellite internet services like Starlink provide reliable connectivity in remote areas, enabling real-time data sharing and communication.

Satellite phones are crucial for maintaining communication in regions without cellular coverage.

4x4 camper vans, like the Field Van Ford Classic 4x4, offer off-road capabilities and customizable sleeping arrangements for field teams.

All-terrain vehicles (ATVs) are essential for accessing areas that larger vehicles cannot navigate.

High-quality extreme weather tents and sleeping bags–Designed for durability and comfort during long expeditions.

Drones with extended flight time and equipped with HD cameras and thermal imaging for aerial surveillance of large, remote areas. Fixed-wing drones or blimp-type UAVs are useful for prolonged, stationary surveillance over specific areas.

Trail cameras with Wi-Fi allow for remote monitoring and instant access to footage.

DNA/eDNA collection Kits–Essential for collecting biological samples that might provide genetic evidence.

DNA samples and storage can be easily taken and preserved using affordable DNA preservation kits that ensure proper storage and transport of collected material.

Night vision binoculars enhance nocturnal observations in dense forested areas.

Portable power stations act as reliable energy sources for electronics during extended field operations (e.g., Jackery or Bluetti systems).

Comprehensive First Aid kits for addressing potential injuries or health issues in remote locations.

Navigation Tools such as GPS devices, maps, and compasses ensure safe navigation in unfamiliar terrain.

Field guides and notebooks for documenting observations, notes, and sketches.

Field support teams should be equipped with heavy-gauge weapons to protect against potential threats from bears or other predators—not for hunting Bigfoot. While bear spray and loud noises might work in some scenarios, they don't guarantee safety. Teams must prioritize preparation to ensure the safety of everyone involved.

While dreaming of billionaire-funded expeditions is fun, most Bigfoot researchers operate with limited resources. In such cases, the key is to make the best use of what's available. Here are some practical tips for independent researchers:

Use the highest-quality equipment within your budget.

Also, remember to hold your phone horizontally while recording to capture clear, stable footage that reflects a more encompassing view of the subject you are recording.

Invest in inexpensive gimbals to reduce camera shake.

Adopt proper protocols for DNA collection and always wear powder-free nitrile gloves to avoid contamination.

Coordinate with other teams to prevent false positives from overlapping research activities.

The best thing any Bigfoot investigator can do is ensure they are alone in their designated research area to minimize the risk of human interference. Meticulous planning, combined with the latest technology, is essential for collecting reliable evidence. At the same time, researchers should never lose sight of the joy of exploring the great outdoors with friends and family.

In the next chapter, we'll explore how data collection tools and predictive modeling can help researchers identify potential hotspots for Bigfoot sightings and analyze historical trends to refine their search strategies.

Chapter 23

The legend of Bigfoot has evolved beyond mere folklore, becoming a significant economic and cultural phenomenon. From tourism-driven events to Bigfoot-branded merchandise, the creature's influence has rippled through local economies and global markets alike. In this chapter, we'll delve into the surprising ways Bigfoot has contributed to industries ranging from tourism to entertainment, showcasing its role as more than just a mythical figure—it's a thriving business.

Bigfoot has become an iconic symbol for many towns and regions across North America, particularly those in areas where sightings have been reported. Communities such as Willow Creek, California, and Whitehall, New York, have embraced their connection to the Sasquatch legend, transforming it into an economic driver.

Festivals and Events: Annual Bigfoot festivals draw thousands of visitors. Events feature guest speakers, sighting reenactments, themed races, and cryptid merchandise. The Willow Creek Bigfoot Festival, for instance, combines fun and folklore with an economic boost to local businesses.

Tours and Expeditions: Bigfoot-themed tours and wilderness expeditions have become increasingly popular.

Guides take enthusiasts into reported hotspot areas, sharing local stories and teaching wilderness tracking techniques. For those seeking a more immersive experience, some tours even simulate Bigfoot encounters, complete with eerie sounds and mysterious footprints.

Museums and Exhibits: Towns like Felton, California, and Sutton, West Virginia, boast dedicated

Bigfoot museums showcasing artifacts, alleged evidence, and historical accounts. These institutions not only preserve cryptozoological lore but also generate tourism revenue. Boring, Oregon, is home to the North American Bigfoot Center (NABC), founded by Cliff Barackman. Personally, I have purchased a few items from the NABC, with which I am very satisfied.

I'm trying to justify (with my wife) buying some of the footprint cast copies Cliff sells at the NABC.

Chapter 24

The commercial appeal of Bigfoot is undeniable. The creature lends its image to countless products and marketing campaigns, from quirky collectibles to major advertising campaigns. Bigfoot's influence has permeated consumer culture.

Branded Products: Bigfoot-branded merchandise includes apparel, mugs, toys, and home décor. Companies capitalize on the creature's mystique, using it as a symbol of adventure and individuality. Popular items often feature phrases like "Gone Squatchin'" or humorous depictions of Sasquatch in everyday situations.

Corporate Campaigns: Bigfoot has been a staple in advertising, notably for Jack Link's Beef Jerky. Their "Messin' with Sasquatch" campaign cleverly mixes humor and intrigue, featuring a hapless Bigfoot being pranked by humans.

Books and Media: The market for Bigfoot-themed literature and films remains robust, encompassing everything from children's books to documentaries and even romance novels. This growing niche has established a dedicated audience eager for new interpretations of the legend.

Regions associated with Bigfoot sightings often see a surge in economic activity tied directly to the creature's lore. Businesses from motels to restaurants thrive during peak tourist seasons, while local artisans and vendors' profit from selling Sasquatch-themed goods.

Areas like the Pacific Northwest promote "Bigfoot Trails," encouraging travelers to explore multiple Sasquatch-related landmarks. These trails often feature stops at scenic viewpoints, alleged sighting locations, and souvenir shops.

Many establishments embrace Bigfoot in their branding, offering dishes like the "Sasquatch Burger" or "Yeti IPA." These establishments cater to tourists and locals alike, blending the Bigfoot mystique with regional flavors.

Beyond its financial contributions, Bigfoot has left an indelible mark on culture, influencing everything from art to identity. It has become a symbol of mystery, resilience, and the enduring human desire to believe in the unknown.

In towns where Bigfoot sightings are part of the lore, the creature is more than just a tourist attraction—it's a point of pride. Residents often adopt Sasquatch as a mascot, reinforcing community ties and regional uniqueness.

Artistic Inspiration: Bigfoot appears in sculptures, murals, and music, reflecting its role as a cultural muse. Artists frequently depict the creature as a guardian of nature, reinforcing themes of environmental conservation.

The Myth and Modernity: In an age dominated by science and technology, Bigfoot represents the possibility that not everything has been discovered or explained. This tension between belief and skepticism keeps the legend alive, bridging the gap between folklore and modern life.

While Bigfoot-related tourism and merchandise can benefit local economies, there are challenges associated with the commercialization of the legend.

Over-tourism can strain small communities, while the emphasis on consumerism risks overshadowing the cultural and folkloric significance of the creature.

To maintain a sustainable balance, many communities are investing in responsible tourism initiatives, such as eco-friendly tours and educational programs that emphasize Bigfoot's role as a symbol of wilderness preservation.

The transformation of Bigfoot from cryptid to cultural icon has created a thriving industry that bridges the gap between myth and commerce. Whether it's a family visiting a Sasquatch museum, a hiker embarking on a Bigfoot expedition, or a fan collecting Bigfoot memorabilia, the creature continues to captivate and inspire.

As the legend of Bigfoot evolves, its economic and cultural impact will likely grow, proving that even a creature as elusive as Sasquatch can leave a tangible footprint on society.

Chapter 25

The search for Bigfoot is more than just a quest for an elusive creature—it is a story that reflects humanity's enduring fascination with the unknown. From the dense forests of the Pacific Northwest to the high mountains of the Himalayas, tales of mysterious beings have captured the imagination of generations. But as the legend of Bigfoot has evolved, so too have the methods and motivations of those who dedicate themselves to the search.

In this chapter, we will reflect on the broader significance of Bigfoot in culture, science, and human curiosity. By examining how the legend has shaped and been shaped by society, we can better understand why Bigfoot endures as a powerful symbol of mystery and discovery.

Few cryptids have achieved the level of cultural recognition that Bigfoot enjoys. From blockbuster movies and television shows to memes and marketing campaigns, Bigfoot has transcended its origins in folklore to become a global icon. This transformation reflects society's ability to adapt myths to new contexts, keeping them relevant across generations.

Bigfoot's appeal lies in its flexibility. To some, the creature represents a primal force of nature, untamed and untouched by civilization. To others, it is a symbol of the unknown, a reminder that the world still holds secrets waiting to be discovered. This duality allows Bigfoot to resonate with a diverse audience, from skeptics who see it as a challenge to science to believers who view it as a guardian of the wilderness.

Bigfoot occupies a unique space at the intersection of science and mythology. On one hand, it is a subject of serious investigation for cryptozoologists, who use scientific tools and methods to search for evidence. On the other hand, it is a creature of legend, its image shaped by centuries of storytelling and cultural memory.

This dual identity can be both a strength and a challenge. For researchers, Bigfoot's mythical status can complicate efforts to gain scientific legitimacy.

Hoaxes, sensationalized media, and a lack of physical evidence have contributed to skepticism within the scientific community. Yet, the same cultural resonance that creates these challenges also fuels the search, inspiring new generations to ask questions and seek answers.

Bigfoot's enduring appeal is deeply rooted in human psychology. The creature represents the tension between what is known and what is unknown, offering a tantalizing glimpse into a world that exists beyond the boundaries of our understanding.

For many, the idea of Bigfoot is comforting. It suggests that the world is not entirely mapped or explained, that there are still places where nature reigns supreme, and mysteries remain unsolved. In a world increasingly dominated by technology and instant access to information, Bigfoot offers a reminder of the wild and the unexplored.

Bigfoot is also a symbol of the natural world and the importance of preserving it. The creatures reported habitats—remote forests, mountain ranges, and swamps—are among the most biodiverse and fragile ecosystems on Earth. For many enthusiasts, the search for Bigfoot is intertwined with a commitment to conservation, a recognition that protecting these wild places is essential not just for potential cryptids but for countless known species as well.

This connection between Bigfoot and the environment has led some researchers to view the creature as an ambassador for wilderness conservation.

By sparking curiosity and drawing attention to remote landscapes, the legend of Bigfoot can inspire a deeper appreciation for nature and a commitment to preserving it for future generations.

As technology continues to evolve, so too will the methods used to investigate Bigfoot. Drones, eDNA sampling, and advanced data analysis offer new tools for researchers, while the rise of citizen science and online communities has democratized the search.

Yet, even as the tools and techniques change, the core of the search remains the same: a desire to explore, to understand, and to connect with the unknown. Whether or not Bigfoot is ever found, the journey itself has value, inspiring curiosity, fostering community, and challenging us to look at the world with fresh eyes.

Bigfoot is more than just a creature of legend; it is a mirror reflecting humanity's deepest hopes, fears, and desires. It represents the possibility that the world is still full of surprises and that the boundaries of human knowledge are not yet fixed.

As we continue to search for Bigfoot, we are also searching for something within ourselves—a sense of wonder, a connection to nature, and a belief in the extraordinary. Whether in the dense forests of the Pacific Northwest or the quiet corners of our imagination,

Bigfoot reminds us that the greatest mysteries are often the ones we carry with us.

Chapter 26

The search for Bigfoot is more than a singular quest—it has become a shared journey that brings together scientists, enthusiasts, and storytellers alike. This chapter reflects on the community of people drawn to the mystery of Bigfoot and how their collective efforts and shared passion have shaped the legend.

Bigfoot research is unique in its ability to bridge diverse perspectives and backgrounds. Scientists, amateur researchers, hikers, hunters, and even casual believers all play a role in keeping the search alive. Each participant brings a different lens to the investigation, creating a tapestry of expertise, curiosity, and creativity that fuels the field.

The advent of online communities has only strengthened this sense of connection. Forums, social media groups, and dedicated websites allow enthusiasts to share sightings, discuss theories, and collaborate on research projects. These platforms provide a space for individuals to engage with the mystery, no matter their location or level of experience.

At the heart of this community is a shared belief in the value of exploration. Whether motivated by scientific curiosity, a love of nature, or the thrill of the unknown, those who pursue Bigfoot are united by a desire to push the boundaries of what we know.

A common thread among Bigfoot enthusiasts is the power of personal encounters. Whether a fleeting glimpse of a shadowy figure in the woods, the sound of unexplained howls at night, or the discovery of massive footprints in soft earth, these experiences leave an indelible mark on those who have them.

These encounters often serve as the catalyst for a lifelong fascination with Bigfoot. For many, the moment they feel they have come close to the creature is enough to ignite a passion for understanding it. Stories of such encounters are shared within the community, inspiring others to venture into the wilderness and seek their own experiences.

While the Bigfoot community is united by a common goal, it is not without its challenges. Differences in methodology, competing theories, and occasional skepticism within the group can create tension. Some researchers focus on scientific approaches, seeking tangible evidence such as DNA or video footage, while others lean into the folkloric and spiritual aspects of the legend.

These differing perspectives can lead to debates over what constitutes credible evidence and how best to conduct research. However, these challenges also highlight the diversity of thought within the community and the importance of open dialogue in advancing the search for Bigfoot.

Storytelling plays a central role in the Bigfoot phenomenon. From firsthand accounts of sightings to speculative theories about the creature's behavior and origins, the stories shared within the community shape the way Bigfoot is perceived.

These narratives are not just about the creature itself—they are about the people who tell them. Each story reflects the hopes, fears, and beliefs of the storyteller, adding a deeply personal dimension to the legend. Through storytelling, the mystery of Bigfoot becomes a shared experience, one that connects people across generations and cultures.

The importance of storytelling extends beyond the Bigfoot community. By sharing their experiences with the broader public, enthusiasts keep the legend alive and inspire others to engage with the mystery. Documentaries, books, podcasts, and even campfire tales ensure that Bigfoot remains a vibrant part of our collective imagination.

One of the most remarkable aspects of Bigfoot research is the diversity of people it attracts. From seasoned scientists to casual hikers, the search for Bigfoot transcends age, background, and geography.

This inclusivity is a testament to the universal appeal of the mystery. Whether motivated by science, folklore, or a love of adventure, everyone who participates in the search contributes to the larger story of Bigfoot. This diversity enriches the field, bringing new ideas, perspectives, and energy to the quest.

At its core, the search for Bigfoot is a shared journey. It is not just about finding evidence or proving the creature's existence—it is about the connections made along the way. For many, the friendships formed and the adventures shared are as meaningful as the search itself.

The Bigfoot community is a reminder that exploration is not a solitary endeavor. It is a collective effort, one that draws strength from the passion and dedication of those involved. Together, they keep the spirit of discovery alive, ensuring that the legend of Bigfoot continues to inspire curiosity and wonder.

As we reflect on the Bigfoot phenomenon, it is clear that the creature represents more than just a mystery of the natural world. It is a symbol of human connection, curiosity, and the enduring power of storytelling.

The Bigfoot community embodies these values, bringing people together in pursuit of a shared dream. Whether through scientific research, personal encounters, or the simple act of sharing a story, each participant adds to the legacy of Bigfoot, ensuring that the legend will continue to captivate and inspire for generations to come.

Chapter 27

The enduring mystery of Bigfoot has inspired countless people to take up the mantle of investigator, storyteller, and explorer. As this manuscript concludes, it's important to reflect on what the legend of Bigfoot reveals—not just about the creature itself, but about human nature and our relationship with the unknown.

Bigfoot is more than just a cryptid; it is a symbol of the wilderness, of mystery, and of our collective longing for discovery. In a world where so much has been mapped, cataloged, and explained, the idea that something as large and intelligent as Bigfoot could still elude us is a powerful reminder of the limits of human knowledge.

For many, Bigfoot represents the untamed spirit of the natural world—a creature that exists beyond the reach of civilization, embodying the wildness that remains in our increasingly developed planet. This connection to the wilderness resonates deeply, inspiring a sense of awe and reverence for the natural world.

The appeal of Bigfoot also lies in its reflection of human curiosity and imagination. The search for Bigfoot is not just a quest for evidence—it is a journey of self-discovery, an exploration of what it means to confront the unknown.

Throughout history, humans have been drawn to mysteries, from the depths of the ocean to the vastness of space. Bigfoot is a terrestrial counterpart to these grand mysteries, a reminder that there are still secrets to uncover on our own planet. The creature taps into a primal part of the human psyche, evoking both fear and wonder in equal measure.

The search for Bigfoot offers lessons that extend beyond cryptozoology. It teaches us the value of perseverance, the importance of critical thinking, and the need for collaboration in the face of uncertainty.

For researchers and enthusiasts, the search is as much about the journey as the destination. The friendships formed, the adventures undertaken, and the insights gained along the way are as valuable as any physical evidence. This sense of shared purpose and exploration is a testament to the enduring power of the human spirit.

As technology advances and new tools become available, the search for Bigfoot will continue to evolve. Drones, eDNA, and artificial intelligence hold the promise of uncovering new evidence, while the global community of researchers ensures that the search remains active and collaborative.

But even as the field grows more sophisticated, the core of Bigfoot research will remain unchanged: the desire to connect with the unknown, to push the boundaries of understanding, and to preserve the mysteries that make life extraordinary.

Whether or not Bigfoot is ever proven to exist, its impact on our culture, our imaginations, and our sense of wonder is undeniable. The creature serves as a reminder that the world is still full of mysteries waiting to be explored, and that the human spirit thrives on the pursuit of discovery.

In the end, Bigfoot is not just a legend—it is a symbol of hope, curiosity, and the endless possibilities that lie just beyond the edge of the known. As long as there are forests to explore and questions to ask, the search for Bigfoot will endure, inspiring generations to come.

Chapter 28

For those intrigued by the pursuit of Bigfoot, access to historical sightings and predictive tools can be invaluable. These resources not only provide insight into past encounters but also help researchers focus their efforts on areas where sightings are more likely to occur. In this chapter, we'll highlight some of the most helpful databases and tools available, as well as the importance of leveraging technology and data in Bigfoot research.

The BFRO maintains one of the most comprehensive and credible databases of Bigfoot sightings across the United States and Canada. Their searchable database allows users to explore reports by state, date, and classification. Detailed accounts include specific locations, witness descriptions, and environmental conditions. Website: https://bfro.net/gdb

Crowdsourcing has become a game-changer for Bigfoot research. With the rise of smartphones and mobile apps, anyone can document their experiences and contribute to the broader effort to uncover the truth. Apps like "iNaturalist," which allow users to record wildlife sightings and share them with scientists, could be adapted specifically for cryptid research.

Imagine an app where users can log sightings, upload photos or videos, and describe the conditions of their encounter. These reports could then be analyzed in real time by experts, creating a centralized and constantly updated database of potential evidence.

The success of crowdsourcing depends on the participation of the public and the ability of researchers to vet and analyze the data. By fostering collaboration between professionals and enthusiasts, the cryptozoological community can create a more comprehensive picture of Bigfoot's possible existence.

While the pursuit of Bigfoot can be exciting, it's essential to prioritize safety and enjoyment. Whether you are hiking in dense forests or setting up trail cameras in remote areas, preparation is key. Always bring the necessary equipment, inform others of your plans, and respect the wilderness.

At the same time, remember that Bigfoot research should be enjoyable. For many enthusiasts, the thrill of exploring nature and searching for the unknown is just as rewarding as the possibility of finding definitive evidence. By approaching the field with a sense of wonder and curiosity, you can make the experience memorable, regardless of the outcome.

Bigfoot research thrives on the passion and dedication of those willing to delve into the unknown. By leveraging historical data, predictive tools, and crowdsourced reports, researchers can refine their strategies and improve their chances of success.

As always, remember to approach the search with a sense of adventure, safety, and respect for the natural world. The truth about Bigfoot may still be out there, waiting to be discovered—but the journey to uncover it is an adventure in its own right.

Next is an alphabetized list of names for Bigfoot and related cryptids, along with their cultural or regional origins. This appendix provides a comprehensive reference for readers interested in the diverse ways this legendary creature has been perceived around the world.

A

- **Almas**: Central Asia, particularly Mongolia and the Caucasus.

- **Arulataq**: Yup'ik people of Alaska.

- **Aswang**: Philippines (shape-shifting creature, sometimes likened to Bigfoot).

B

- **Bagwajiwinini**: Ojibwa/Anishinaabe First Nations.

- **Bigfoot**: Modern North American name, popularized in the 20th century.

- **Bukwas**: Kwakwaka'wakw First Nations of British Columbia (a forest spirit often compared to Bigfoot).

D

- **Didi**: Georgian folklore (Caucasus region).

- **Dsonoqua**: Nuu-chah-nulth and other Pacific Northwest Coast First Nations.

E

- **Ebu Gogo**: Flores Island, Indonesia (small humanoid cryptid but sometimes linked to Bigfoot lore).

F

- **Fouke Monster**: Arkansas, U.S. (associated with the Boggy Creek Monster legend).

G

- **Gigantopithecus**: Scientific theory often linked to Bigfoot (extinct giant ape species).

- **Grassman**: Ohio, U.S. (Bigfoot-like creature that purportedly builds grass nests).

H

- **Harry Man**: Texas, U.S. (often tied to local legends of a wild, human-like figure).

- **Hibagon**: Japan (Bigfoot-like creature).

- **Hidebehind**: U.S. folklore (shy forest creature sometimes compared to Bigfoot).

K

- **Kushtaka**: Tlingit people of Alaska (a shape-shifting creature resembling Bigfoot).

L

Lofa: Liberia, West Africa (a large, hairy cryptid reportedly sighted near the Lofa River).

M

- **Mande Barung**: Meghalaya, India (Bigfoot-like creature).

- **Mapinguari**: Amazonian rainforest, Brazil (giant sloth-like cryptid but often likened to Bigfoot).

- **Mo-Mo the Monster**: Missouri, U.S. (a Bigfoot-like creature seen in the 1970s).

N

- **Nantiinaq**: Alaskan folklore, particularly among the Dena'ina Athabaskans.

- **Nyalmo**: Nepal (a giant, hairy humanoid similar to the Yeti).

O

- **Orang Gadang**: Malaysia (literally "big man").

- **Orang Pendek**: Sumatra, Indonesia (a small, bipedal primate-like creature often linked to Bigfoot lore).

R

- **Rock Ape**: Vietnam (creature described by soldiers during the Vietnam War, often linked to Bigfoot narratives).

S

- **Sasquatch**: Coast Salish First Nations of the Pacific Northwest (the most widely recognized indigenous term for Bigfoot).

- **Shampe**: Choctaw Nation (a Bigfoot-like creature).

- **Skunk Ape**: Florida and Southeastern U.S. (a Bigfoot-like creature known for its foul odor).

- **Stiyaha**: Miwok people of California (a nocturnal creature resembling Bigfoot).

T

- **Ts'emekwes**: Lummi people of the Pacific Northwest.

- **Tsul 'Kalu**: Cherokee Nation (a Bigfoot-like spirit tied to mountain regions).

W

- **Wendigo**: Algonquin and other Indigenous cultures of Canada (a mythical spirit, sometimes conflated with Bigfoot-like entities).

- **Wildman**: European and U.S. folklore (general term for a wild, human-like creature).

- **Wooly Booger**: Southern U.S. (a humorous or folkloric term for Bigfoot).

- **Wood Booger**: Appalachian U.S. (a colloquial term for Bigfoot or similar creatures).

- **Wudewasa**: Old English/Medieval Europe (mythical forest dwellers, sometimes likened to Bigfoot).

Y

- **Yeren**: China (a wild man or ape-man in Hubei Province).

- **Yeti**: Himalayas (snow-dwelling humanoid cryptid, also known as the "Abominable Snowman").

- **Yowie**: Australia (Bigfoot-like creature from Aboriginal folklore).

NOTES:

www.ingramcontent.com/pod-product-compliance
Lightning Source LLC
LaVergne TN
LVHW052313200926
848481LV00002B/550